Wolf von Wahl

The Equations of Navier-Stokes and Abstract Parabolic Equations

Aspects of Mathematics
Aspekte der Mathematik

Herausgeber Klas Diederich

Vol. E1: G. Hector/U. Hirsch, Introduction to the Geometry of Foliations, Part A

Vol. E2: M. Knebusch/M. Kolster, Wittrings

Vol. E3: G. Hector/U. Hirsch, Introduction to the Geometry of Foliations, Part B

Vol. E4: M. Laska, Elliptic Curves over Number Fields with Prescribed Reduction Type

Vol. E5: P. Stiller, Automorphic Forms and the Picard Number of an Elliptic Surface

Vol. E6: G. Faltings/G. Wüstholz et al., Rational Points
(A Publication of the Max-Planck-Institut für Mathematik, Bonn)

Vol. E7: W. Stoll, Value Distribution Theory for Meromorphic Maps

Vol. E8: W. von Wahl, The Equations of Navier-Stokes and Abstract Parabolic Equations

---

Band D1: H. Kraft, Geometrische Methoden in der Invariantentheorie

Die in dieser Reihe veröffentlichten Texte wenden sich an graduierte Studenten und alle Mathematiker, die ein aktuelles Spezialgebiet der Mathematik neu kennenlernen wollen, um Ergebnisse und Methoden in der eigenen Forschung zu verwenden oder um sich einfach ein genaueres Bild des betreffenden Gebietes zu machen. Sie sollen eine lebendige Einführung in forschungsnahe Teilgebiete geben und den Leser auf die Lektüre von Originalarbeiten vorbereiten.
Die Reihe umfaßt zwei Unterreihen, eine deutsch- und eine englischsprachige.

Wolf von Wahl

The Equations of Navier-Stokes and Abstract Parabolic Equations

Springer Fachmedien Wiesbaden GmbH

Prof. Dr. *Wolf von Wahl* is Professor of Applied Mathematics at the University of Bayreuth, Fed. Rep. of Germany.

AMS Subject Classification: 35 Q 10, 35 K 22

1985

Originally published by Friedr. Vieweg & Sohn Verlagsgesellschaft mbH, Braunschweig in 1985.

Produced by Lengericher Handelsdruckerei, Lengerich

ISBN 978-3-528-08915-3 ISBN 978-3-663-13911-9 (eBook)
DOI 10.1007/978-3-663-13911-9

Contents

0. Introduction, Auxiliary Propositions, and Notations

§ 1. Introduction

This book is mainly devoted to the study of the initial boundary value problem for the system of Navier-Stokes

$$(0.1.1)\quad \begin{cases} \dfrac{\partial u}{\partial t} - \nu\Delta u + u\cdot\nabla u + \nabla\pi = f, \\ \nabla\cdot u = 0, \end{cases}$$

$$(0.1.2)\quad u(t,x) = 0,\ t > 0,\ x \in \partial\Omega,$$

$$(0.1.3)\quad u(0,x) = \varphi(x)$$

over a cylindrical domain $(0,T)\times\Omega \subset \mathbb{R}^{n+1}$. This system describes the velocity u and pressure π of a viscous incompressible fluid under the influence of an external force f. The viscosity ν is assumed to be constant. Of physical importance are only the cases $n = 3$ and $n = 2$ (if the data depend on two variables only). Because of its mathematical interest however, we intend to develop a theory for arbitrary n. In most cases the theory available for $n = 3$ can be carried over to any $n \geq 3$ (after suitable modifications). Ω is the space domain filled out by the fluid; it is assumed to be bounded although many of our results remain valid for the cases that Ω is the whole of $\mathbb{R}^n$ or an exterior domain.

The access to (0.1.1) is a functional analytic one. Therefore we treat in great detail the local strong solvability of abstract parabolic equations

$$(0.1.4) \begin{cases} u' + Au + M(u) = 0, \\ u(0) = \varphi \end{cases}$$

in a Banach space B. Here $-A$ generates an analytic semigroup in B and M is a nonlinear mapping fulfilling suitable Lipschitz conditions. In particular we deal with the question which initial data are admissible in order to guarantee the unique existence of a sufficiently "nice" solution in some maximal interval $(0,T(\varphi))$. Also the question how to characterise the case $T(\varphi) < +\infty$ is studied in detail. This theory is then applied to (0.1.1) by making use of the well known fact that (0.1.1) can be transformed into an equation like (0.1.4) if we apply formally the projection P_p of some $(L^p(\Omega))^n$ onto its divergence free part $H_p(\Omega)$. Then we get

$$u' - \nu P_p \Delta u + P_p(u \cdot \nabla u) = P_p f,$$
$$u(0) = \varphi,$$

where φ is assumed to be divergence free in some generalized sense. The negative Stokes-operator $-A = \nu P_p \Delta$ turns out to be the generator of an analytic semigroup in the Banach space $H_p(\Omega)$ and $M(u) = P_p(u \cdot \nabla u)$ fulfills Lipschitz conditions being covered by our abstract theory; of course we have assumed that, because of $\nabla \cdot u(t,x) = 0$ in (0.1.1), the function u is invariant under P_p.

As it is well known, besides this abstract setting one can construct a possibly non-unique and non-regular weak solution to (0.1.1) for all times and any space dimension n; this result goes back to E. Hopf [Ho]. Following the ideas of Leray [Ler] and Serrin [Ser2] one main concern of this book is to study the connections between local (in t) strong solutions and weak solutions. Thus we get in particular results concerning the possible singularities (in t) of weak solutions.

The second aim of the book is to study criteria for the uniqueness and global (in t) regularity of weak solutions. For that purpose we have not only to use our abstract theory but also to rely on a fundamental idea of Serrin [Ser1, Ser2]: Serrin has studied the consequences of the assumption that a weak solution u of (0.1.1) is in some space $L^s((0,T),(L^r(\Omega))^n)$ for suitable s,r. For that purpose he has introduced a "critical quantity" $q(s,r) = \frac{2}{s}+\frac{n}{r}$. We deal in detail with q(s,r) and, in particular, we try to cover all marginal values of q(s,r) which were not considered by Serrin.

In order to get results as best as possible for the global regularity it is not surprising that we need so called maximal regularity results for the linear part of (0.1.1). The estimates being necessary for that aim are not provided by the theory of semigroups but by the potential theoretical estimates of Solonnikov [Sol 1,2]. A consequence of these estimates is e.g. the fact that $-A = \nu P_p \Delta$ generates an analytic semigroup in $H_p(\Omega)$.

In general one can say that the best results are furnished by an interaction between the various methods to solve (0.1.1). In chapter V we use abstract methods and estimates for the concrete equations to prove both, the global existence of solutions to certain parabolic systems and the regularity of weak solutions of the Navier-Stokes equations being in $C^0([0,T], (L^n(\Omega))^n)$. The abstract method employed here for equations like (0.1.4) gives a very precise description of what happens when the length of the maximal interval of existence is finite; this enables us to study the question of global existence for semilinear parabolic systems

$$u^{l\prime} + \sum_{\substack{|\alpha|\leq m,\\ |\beta|\leq m}} D^\alpha (A_{\alpha\beta}(x) D^\beta u^l) + f^l(t,x,u) = 0, \quad 1 \leq l \leq L,$$

with growth restrictions on f as weak that they could not be treated by previous methods. It is our hope that we have given an impression of that interplay to the interested reader. In order not to overburden the presentation we were forced just to describe some results on questions being treated in more detail in other text books ([Tem1,L],[Li]), as there are: The projection on the divergence free part of $(L^p(\Omega))^n$ (see (O.2)), Solonnikov's potential theoretical estimates (see III.1), the existence of global weak solutions to (O.1.1) (see IV.1).

The reader is supposed to have some knowledge on unbounded operators in a Banach space and Banach space valued functions. The theory of semigroups of bounded operators (including fractional powers of the negative generator) is briefly described in I., whereas the other prerequisites of functional analysis are given in O:2, including Sobolev spaces of integer order and the real and complex interpolation spaces between them. The reader being not familiar with these interpolation spaces may imagine them as intermediate spaces between the Sobolev spaces of integer order; for them the imbedding theorems still are valid if the integer order of differentiation is replaced by a fractional one, provided the other necessary conditions between the fractional order of differentiation and the exponents of integration involved are fulfilled. In particular the complex interpolation spaces are needed to characterise precisely the domains of definition of the fractional powers of the Stokes operator and other elliptic operators.

In each chapter we have added in a seperate paragraph some comments. Besides of giving further references to the literature we also use them to outline other aspects of the theory developed here. Sections within two stars: *...* freely use knowledge on fields having not been touched in this book.

The reader not being interested in the abstract theory being developed in I.,II. but only in the Navier-Stokes equations itself may start directly with III. and have a look to I.,II. only if we refer to these chapters.

We have not intended to give a complete bibliography here; the choice is determined by the degree of connection with the present work and the taste of the author. For further references see [Tem],[L].

I want to take the opportunity to thank my colleague H. Sohr (University of Paderborn) for many helpful comments and for reading the manuscript before it was printed, and Mrs. R. Göttgens for her excellent typing.

§ 2. Auxiliary Propositions and Notations

The reader is supposed to be acquainted with the basic concepts of the theory of bounded and unbounded linear operators A in a Banach space B and a Hilbert space H, including the notion and the basic properties of continuous and analytic semigroups generated by unbounded operators -A. This material is now part of many well known text books, see e.g. [Ta, 1 and 3], [Y2, IX]. Some part of it, however, is briefly reviewed in I.1.

We also assume that the reader is familiar with the basic facts about functions $f: I \to B$, where B is a Banach space with norm $\| \ \|_B$, and I is an interval of the real axis. $C^k(I,B)$, $k \in \mathbb{N} \cup \{0\}$, is the set of k-times continuously differentiable mappings from I into B. $C^{k+\alpha}(I,B)$, $0 < \alpha < 1$, is the subset of $C^k(I,B)$, where

$$\frac{\| \frac{d^k f}{dt^k}(t) - \frac{d^k f}{dt^k}(s) \|_B}{|t-s|^\alpha} \leq C(K),$$

for all compact intervals $K \subseteq I$ and all $t, s \in k$ with $t \neq s$. $C^{0,1}(I,B)$ indicates uniform Lipschitz continuity on every $K \subseteq I$. We frequently need the Bochner integral $\int_I f(s)\, ds$, and $L^p(I,B)$, $1 \leq p \leq +\infty$, is the Banach space of all a.e. defined measurable functions $f: I \to B$ with $\int_I \|f(s)\|_B^p\, ds < +\infty$, $1 \leq p < +\infty$ (its norm being $(\int_I \|f(s)\|_B^p\, ds)^{1/p}$) or with

$$\|f\|_{L^\infty(I,B)} := \operatorname*{ess\,sup}_{s \in I} \|f(s)\|_B < +\infty, \ p = +\infty.$$

The reader can find the necessary material in [Ta, 1], [Y2, V] and, in particular, in [GGZ, Kapitel IV]. In [GGZ, IV, § 1] also the notion of the weak derivative f' is introduced. We simultaneously use f' for the weak derivative, the strong derivative $\frac{df}{dt}$ and the distributional derivative with respect to the (time) variable t if f is a concrete function on a cylinder $(0,T)\times\Omega$ (Ω = open subset of $\mathbb{R}^n$); it is clear from the connection which notion is meant. In most cases the space B is reflexive or even reflexive and separable; then $L^q(I,B^*)$ is the dual space of $L^p(I,B)$ if $1 \leq p < +\infty$, $\frac{1}{q}+\frac{1}{p} = 1$ ($q = +\infty$ is admitted, B* is the dual space of B; cf. [GGZ, IV, § 1]). Besides the derivatives already mentioned we can also define the derivative in the sense of vector valued distributions of an $f:I \to B$ with $f \in L^1(K,B)$ for every compact interval $K \subseteq I$. This derivative is also denoted by f'; if the weak derivative f' having already been introduced exists and is in $L^1(I,B)$, then this weak derivative is also the derivative of f in the sense of vector valued distributions.

Let B,B' be two Banach spaces with $B \subset B'$ with a continuous imbedding. Let, moreover, B be dense in B'. Let $W_{p,q} = \{u | u:I \to B,\ u':I' \to B',\ u \in L^p(I,B),\ u' \in L^q(I,B')\}$ for some q,p, $1 < q,p < +\infty$ and some bounded interval I. $W_{p,q}$ is a Banach space with norm $\|u\|_{W_{p,q}} = (\int_I \|u(\sigma)\|_B^p\, d\sigma)^{1/p} + (\int_I \|u'(\sigma)\|_{B'}^q\, d\sigma)^{1/q}$. Each $u \in W_{p,q}$ can be approximated by elements

$$(0.2.1)\quad u_n(t) = \sum_{\nu=1}^{N_n} \zeta_\nu^{(n)}(t)v_\nu^{(n)}$$

in such a way that $u_n \to u$ in $L^p(I,B)$, $u_n' \to u'$ in $L^q(I,B')$; here $\zeta_\nu^{(n)}:\bar{I} \to \mathbb{C}$ are of class C^1 and the $v_\nu^{(n)}$ are elements of B; we can also prescribe $u_n(a) = u(a)$ or $u_n(b) = u(b)$ if a,b are the endpoints of I (for a proof in a similar situation see [GGZ, IV, § 1]); observe that $u \in W_{p,q}$ furnishes the relation $u \in C^0(\bar{I},B')$, and therefore it makes sense to write down u(a), u(b). Often the following situation occurs: B' is a Hilbert space with scalar product (.,.). Then we have

$$B \subset B' \subset B^*$$

with continuous imbeddings, moreover B' is dense in B* and the scalar product between B and B* can be denoted by (.,.) as the scalar product in B' ([GGZ, IV, § 1]). Let p, p_o be given with $1 < p \leq p_o < +\infty$. We define q, q_o by $\frac{1}{p}+\frac{1}{q} = \frac{1}{p_o}+\frac{1}{q_o} = 1$, and we set

$$(0.2.2)\quad X = L^p(I,B) \cap L^{p_o}(I,B')$$

with norm

$$(0.2.3)\quad \|u\|_X = \|u\|_{L^p(I,B)} + \|u\|_{L^{p_o}(I,B')} .$$

Then the dual space X* is $L^q(I,B^*) + L^{q_o}(I,B')$ with norm

$$(0.2.4)\quad \|f\|_{X^*} = \inf_{\substack{f_1 \in L^q(I,B^*),\\ f_2 \in L^{q_o}(I,B'),\\ f_1+f_2=f}} \max\left(\|f_1\|_{L^q(I,B^*)}, \|f_2\|_{L^{q_o}(I,B')}\right).$$

Thus X is reflexive. For $u \in X$ we denote by u' its derivative in the sense of vector valued distributions, and we set

$$W_{p,q,p_o,q_o}(B,B',I) = W = \{u \mid u \in X,\ u' \in X^*\}.$$

Thus W becomes a Banach space if we set

$$(0.2.5)\quad \|u\|_W = \|u\|_X + \|u'\|_{X^*}.$$

We have $W \subset C^{0}(I,B^{*})$, $W \subset C^{0}(I,B')$; the inclusions are continuous imbeddings if I is compact. Moreover we have the following formula of partial integration

$$(0.2.6)\quad (u(t),v(t)) - (u(s),v(s)) =$$

$$= \int_{s}^{t} \{(u'(\tau),v(\tau))+(u(\tau),v'(\tau))\}\, d\tau,$$

$$s,t \in I,\ u,v \in W.$$

Consequently

$$(0.2.7)\quad \|u(t)\|_{B'}^{2} - \|u(s)\|_{B'}^{2} = 2\,\mathrm{Re}\int_{s}^{t} (u'(\tau),u(\tau))\, d\tau.$$

Moreover $C^{1}(I,B) \cap W$ is dense in W. Cf. [GGZ, IV, § 1] for this material.

Let Ω be a bounded open subset of $\mathbb{R}^{n}$. Its closure is denoted by $\bar{\Omega}$, its boundary by $\partial\Omega$.

Functions and distributions will be complex-valued, unless otherwise stated explicitly. Let $1 \leq p \leq \infty$. The collection of all measurable functions for which the pth power of the modulus is integrable in Ω is denoted by $L^{p}(\Omega)$. $L^{p}(\Omega)$ for $p = \infty$ is the collection of all functions which are essentially bounded and measurable in Ω. The norm of a function u in $L^{p}(\Omega)$ is denoted by $\|u\|_{L^{p}(\Omega)}$ or $\|.\|_{p}$. Let m be a positive integer. $H^{m,p}(\Omega) = W^{m,p}(\Omega)$ consists of all functions which, together with their derivatives to order m in the sense of distributions, belong to $L^{p}(\Omega)$. The norm of a function $u \in H^{m,p}(\Omega)$ is defined by

$$\begin{cases} \|u\|_{H^{m,p}(\Omega)} = \|u\|_{m,p} = \left(\sum_{|\alpha| \le m} \|D^\alpha u\|_p^p \right)^{1/p}, \quad 1 \le p < \infty, \\ \|u\|_{H^{m,\infty}(\Omega)} = \|u\|_{m,\infty} = \max_{|\alpha| \le m} \|D^\alpha u\|_\infty \end{cases}$$

here α is an n-tuple of non-negative integers $(\alpha_1,\ldots,\alpha_n)$ and we have designated symbols as follows: $|\alpha| = \alpha_1+\ldots+\alpha_n$, $D = (i^{-1}\partial/\partial x_1,\ldots,i^{-1}\partial/\partial x_n)$ and $D^\alpha = (i^{-1}\partial/\partial x_1)^{\alpha_1}\ldots(i^{-1}\partial/\partial x_n)^{\alpha_n}$. $W^{m,p}(\Omega)$ is a Banach space and, especially, $H^{m,2}(\Omega)$ is a Hilbert space. We denote by $C^m(\Omega)$ the collection of all functions which are m-times continuously differentiable in Ω. Among them, the collection of functions whose supports are compact in Ω will be denoted by $C_o^m(\Omega)$. Another set of elements of $C^m(\Omega)$ is $C^m(\overline{\Omega})$. It consists of all functions u for which $D^\alpha u$ can be continued continuously to $\overline{\Omega}$, $|\alpha| \le m$. $C^m(\overline{\Omega})$ thus defined is a Banach space with the norm

$$\|u\|_{C^m(\overline{\Omega})} = \sum_{|\alpha| \le m} \|D^\alpha u\|_{C^o(\overline{\Omega})},$$

$$= \sum_{|\alpha| \le m} \sup_{x \in \Omega} |D^\alpha u(x)|.$$

The closure of $C_o^\infty(\Omega)$ with respect to the norm of $H^{m,p}(\Omega)$ is denoted by $\overset{o}{H}{}^{m,p}(\Omega)$.

Let $\beta \in (0,1)$. $C^{m+\beta}(\Omega)$ is the subset of $C^m(\Omega)$ for which

$$\frac{|D^\alpha u(x) - D^\alpha u(y)|}{|x-y|^\beta} \le C(u,K), \quad |\alpha| = 2m,$$

for every compact set $K \subset \Omega$ and every pair $(x,y) \in K \times K$ with $x \neq y$. $C^{m+\beta}(\overline{\Omega})$ consists of all $u \in C^m(\overline{\Omega})$ with

$$\frac{|D^\alpha u(x)-D^\alpha u(y)|}{|x-y|^\beta} \le C(u), \quad |\alpha| = 2m, \quad x,y \in \Omega, \quad x \neq y.$$

$C^{o,1}(\Omega)$, $C^{o,1}(\overline{\Omega})$ indicate Lipschitz continuity. $C^{m+\beta}(\overline{\Omega})$ is also a Banach space with norm

$$\|u\|_{C^{m+\beta}(\overline{\Omega})} = \|u\|_{C^m(\overline{\Omega})} + \sum_{|\alpha|=m} \sup_{\substack{x,y\in\Omega,\\ x\neq y}} \frac{|D^\alpha u(x)-D^\alpha u(y)|}{|x-y|^\beta}.$$

$(L^p(\Omega))^k = L^p(\Omega) \times \ldots \times L^p(\Omega)$ (k-times), $(H^{m,p}(\Omega))^k$, $(W^{m,p}(\Omega))^k, \ldots, (C^{m+\beta}(\Omega))^k$, k a positive integer, are defined in the usual way. If B is a Banach space then B^k can be made a Banach space too by setting

$$\|x\|_{B^k} = \left(\sum_{i=1}^{k} \|x_i\|^2\right)^{\frac{1}{2}}$$

for $x = (x_1,\ldots,x_k)$.

We will frequently write $C^{m+\alpha}(\Omega)$, $C^{m+\alpha}(\overline{\Omega})$ $(\alpha\in(0,1))$ instead of $C^{m+\beta}(\Omega)$, $C^{m+\beta}(\overline{\Omega})$ if no confusion can arise. Also the norm of $(H^{m,p}(\Omega))^k$ will sometimes be denoted by $\|\cdot\|_{m,p}$. In an obvious way many of our definitions can be carried over to $\mathbb{R}^n$ instead of Ω. We will freely use notions as $L^p(\mathbb{R}^n)$, $H^{m,p}(\mathbb{R}^n)$, $C_o^m(\mathbb{R}^n)$ etc.

Some lemmas on the above-mentioned function spaces will be stated, but the assumptions they make about the smoothness of the boundary $\partial\Omega$, for example, do not necessarily aim at getting the best results. Principally, we follow the work of Browder [Br].

Definition 0.2.1: Let Ω be as above. Ω is said to be of class C^m or more precisely regular of class C^m if, for each point x of the boundary $\partial\Omega$ of Ω, there exists a neighbourhood N and a diffeomorphism Φ from N to $\{y \in \mathbb{R}^n : |y| = (y_1^2+\ldots+y_n^2)^{1/2} < 1\}$ such that $\Phi(N\cap\Omega) = \{y \in \mathbb{R}^n : |y| < 1,\ y_1 > 0\}$ and $\Phi(N\cap\partial\Omega) = \{y \in \mathbb{R}^n : |y| < 1,\ y_1 = 0\}$, and that each component of both Φ and Φ^{-1} is m-times continuously differentiable.

Mostly we will assume that Ω is of class C^∞. The lemma to follow is refered to as "Sobolev's imbedding" or simply "Sobolev".

Lemma 0.2.1: Let Ω be a bounded open subset of $\mathbb{R}^n$ which is regular of class C^{m+1}. Let j denote a non-negative integer and p,r real numbers belonging to $(1,\infty)$.

1. If $0 \le j \le m$ and $p^{-1}-(m-j)n^{-1} \le r^{-1} \le p^{-1}$, then $H^{m,p}(\Omega) \subset H^{j,r}(\Omega)$ and there exists a constant C such that

   $$\|u\|_{j,r} \le C\|u\|_{m,p}^{\lambda}\|u\|_{p}^{1-\lambda}$$

   is valid, where $\lambda = nm^{-1}(p^{-1}-r^{-1}+jn^{-1})$.

2. If $p^{-1} < (m-j)n^{-1}$, then $H^{m,p}(\Omega) \subset C^j(\bar{\Omega})$ and there exists a constant C such that

   $$\|u\|_{C^j(\bar{\Omega})} \le C\|u\|_{m,p}^{\mu}\|u\|_{p}^{1-\mu}$$

   is valid for all $u \in W_p^m(\Omega)$, where $\mu = nm^{-1}p^{-1}+jm^{-1}$.

3. If $0 < m-j-\frac{n}{p} < 1$ then we even have

   $$H^{m,p} \subset C^{j+\alpha}(\bar{\Omega}),\quad 0 \le \alpha \le m-j-\frac{n}{p}.$$

   The imbeddings are compact if in 1. $j < m$, $p^{-1}-(m-j)n^{-1} < r^{-1}$, in 3. $\alpha < m-j-\frac{n}{p}$.

As a special case of 1., we set $r = p$. Then, for $0 \le j < m$ and every $u \in W_p^m(\Omega)$, the following holds:

$$\|u\|_{j,p} \le C\|u\|_{m,p}^{j/m}\|u\|_{p}^{(m-j)/m}.$$

This is called the _interpolation inequality_. But there is a still more general interpolation inequality which we call the Gagliardo-Nirenberg interpolation inequality.

Lemma 0.2.2: Let Ω _be of class_ C^∞. _Let_ u _be any function in_ $H^{m,r}(\Omega) \cap L^q(\Omega)$, $1 \leq r,q \leq +\infty$. _For any integer_ j, $0 \leq j < m$, _and for any number_ a _in the interval_ $j/m \leq a < 1$, _set_

$$\frac{1}{p} = \frac{j}{n} + a\left(\frac{1}{r} - \frac{m}{n}\right) + (1-a)\frac{1}{q}.$$

_If_ $m-j-n/r$ _is not a nonnegative integer, then_

$$(0.2.8) \quad \|D^\alpha u\|_{L^p(\Omega)} \leq c\|u\|^a_{H^{m,r}(\Omega)} \|u\|^{1-a}_{L^q(\Omega)}, \quad |\alpha| = j,$$

_If_ $m-j-n/r$ _is a nonnegative integer, then_ (0.2.8) _holds for any_ $a \in [j/m, 1)$. _The constant_ c _depends on_ Ω, r, q, m, j, a.

For a proof see [F, p. 27]. We want to mention that _(0.2.8) also holds if_ Ω _is replaced by the whole of_ $\mathbb{R}^n$. In this connection we want to make a few remarks on the validity of the chain rule. Let $f:\mathbb{C} \to \mathbb{C}$ be of class C^j for some $j \leq m$, let $u \in H^{m,p}(\Omega)$. Then often $f(u)$ is in $H^{j,q}(\Omega)$, and all derivatives of $f(u)$ can be gained by formal differentiation, i.e. $D^\beta f(u)$ consists of expressions

$$(0.2.9) \quad f^{(\nu)}(u) \prod_{i=1}^{N} D^{\beta_i} u$$

with $\nu \leq j$, $\sum_{i=1}^{N} \beta_i = \beta$, $|\beta| \leq j$; $f^{(\nu)}$ denotes a derivative of f order ν with respect to the real or imaginary part of u. A sufficient condition for this is that the expressions

$$(0.2.10) \quad f^\nu(v_o) \prod_{i=1}^{N} D^\beta v_i$$

are continuous in $L^q(\Omega)$ with respect to $(v_0,v_1,\dots,v_N) \in H^{m,p}(\Omega)\times H^{m,p}(\Omega)\times\dots\times H^{m,p}(\Omega)$. A well known example occurs if $j=m$, $q=p$, $m>\frac{n}{p}$. It is also well known that for a Lipschitz continuous f with bounded derivatives being continuous with the exception of countably many points, the expression f(u) is in $H^{1,q}(\Omega)$ for any $q>1$ if u is so, and the derivatives can be computed by formal differentiation. If e.g. $f(\mathbb{R})\subset\mathbb{R}$ and u is real then

$$(0.2.11)\quad \frac{\partial}{\partial x_i}f(u) = f'(u)\frac{\partial u}{\partial x_i}.$$

If $f(0)=0$ then even $f(\overset{\circ}{H}{}^{1,q}(\Omega))\subset\overset{\circ}{H}{}^{1,q}(\Omega)$. A proof can be found in [Sim].

Besides the spaces $H^{m,p}(\Omega)=W^{m,p}(\Omega)$ we need "intermediate" spaces $H^{s,p}(\Omega)$, $W^{s,p}(\Omega)$. This is also the case with $\overset{\circ}{H}{}^{m,p}(\Omega) = \overset{\circ}{W}{}^{m,p}(\Omega)$. We follow the work of Triebel [Tr1, 2.3, 4.3, 4.9] and define

$$H^{s,p}(\Omega) = [L^p(\Omega),H^{m,p}(\Omega)]_{\Theta=\frac{s}{m}},\quad s\geq 0,$$

as the complex interpolation space between $L^p(\Omega)$ and $H^{m,p}(\Omega)$. This definition does not depend on m. For the real interpolation space

$$W^{s,p} = (L^p(\Omega),W^{m,p}(\Omega))_{\Theta=\frac{s}{m}},\quad s\geq 0,$$

we refer all the same to [Tr1, 2.3,4.3,4.9]. This definition also does not depend on m. Lemma 0.2.1 then also holds for fractional m,j, and for $W^{m,p}(\Omega)$, $W^{j,r}(\Omega)$ instead of $H^{m,p}(\Omega)$, $H^{j,r}(\Omega)$. Comparing $H^{s,p}(\Omega)$ and $W^{s,p}(\Omega)$ we get

$$(0.2.12)\quad H^{s,2}(\Omega) = W^{s,2}(\Omega),$$

$$(0.2.13)\quad H^{s+\varepsilon,p}(\Omega)\subset W^{s,p}(\Omega)\subset H^{s-\varepsilon,p}(\Omega),$$

(0.2.14) $W^{s-\varepsilon,p}(\Omega) \subset H^{s,p}(\Omega) \subset W^{s+\varepsilon,p}(\Omega)$

with continuous imbeddings and for $0 < \varepsilon \leq s$ (cf. [Tr1, 2.3,4.6]). As usual $\overset{o}{H}{}^{s,p}(\Omega), \overset{o}{W}{}^{s,p}(\Omega)$ are the completions of $C_o^\infty(\Omega)$ with respect to the norm of $H^{s,p}(\Omega), W^{s,p}(\Omega)$. Then

(0.2.15) $\overset{o}{H}{}^{s,p}(\Omega) = [L^p(\Omega), \overset{o}{H}{}^{m,p}(\Omega)]_{\theta=\frac{s}{m}}$,

(0.2.16) $\overset{o}{W}{}^{s,p}(\Omega) = (L^p(\Omega), \overset{o}{W}{}^{m,p}(\Omega))_{\theta=\frac{s}{m}}$

but the values s = nonnegative integer $+\frac{1}{p}$ are excluded. E.g. the inclusion $u \in [L^p(\Omega), \overset{o}{H}{}^{1,p}(\Omega)]_{1/p}$ imposes a boundary condition on u, whereas $u \in \overset{o}{H}{}^{1/p,p}(\Omega)$ does not: We have

(0.2.17) $\overset{o}{H}{}^{s,p}(\Omega) = H^{s,p}(\Omega)$, $\overset{o}{W}{}^{s,p}(\Omega) = W^{s,p}(\Omega)$

for $0 \leq s \leq \frac{1}{p}$ (cf. [Tr1, 4.3]).

Of course we have the same results for the spaces $(H^{s,p}(\Omega))^k = H^{s,p}(\Omega) \times \ldots \times H^{s,p}(\Omega)$ (k times), $(W^{s,p}(\Omega))^k, (\overset{o}{H}{}^{s,p}(\Omega))^k, (\overset{o}{W}{}^{s,p}(\Omega))^k$. The norm of $(H^{s,p}(\Omega))^k$ or $(W^{s,p}(\Omega))^k$ sometimes are denoted by $\|\cdot\|_{s,p}$ (if no confusion can arise).

The dual space of $\overset{o}{H}{}^{s,p}(\Omega)$ via the $L^q(\Omega)$-$L^p(\Omega)$ scalar product is denoted by $H^{-s,q}(\Omega)$, $\frac{1}{p}+\frac{1}{q} = 1$, $+\infty > p > 1$, $s \geq 0$. We set

$$H_*^{1,p}(\Omega) = \{u \mid u \in (\overset{o}{H}{}^{1,p}(\Omega))^n,\ \nabla \cdot u = 0\}$$

for $p > 1$, and we denote its dual space by $H_*^{-1,q}(\Omega)$ (via the $L^q(\Omega)$-$L^p(\Omega)$ scalar product). This means that

$$H_*^{-1,q}(\Omega) \supset (H^{-1,q}(\Omega))^n \supset (\overset{o}{H}{}^{1,q}(\Omega))^n \supset \overset{o}{H}{}_*^{1,q}(\Omega)$$

with continuous imbeddings.

$L^p_M(\Omega), \overset{\circ}{H}{}^{1,p}_M(\Omega), C^\infty_{o,M}(\Omega)$ are the subspaces of $L^p(\Omega), \overset{\circ}{H}{}^{1,p}(\Omega)$, $C^\infty_o(\Omega)$ consisting of all functions having mean value 0, $1 < p < +\infty$. As Bogovskij ([Bo]) has stated, the operator div maps $(\overset{\circ}{H}{}^{1,p}(\Omega))^n$ onto $L^p_M(\Omega)$. A proof was given by Erig [Er]. Ω is assumed to be pathwise connected.

Let $p > 1$, $s > \frac{1}{p}$ and $u \in W^{s,p}(\Omega)$, then the restriction $u|\partial\Omega$ of u to the boundary $\partial\Omega$ is defined (cf. [Tr1, 3.6.1.]); in particular, for functions belonging to $C^{[s]+1}(\bar{\Omega})$, this restriction coincides with that in the usual sense. With Ω being regular of class C^∞, the collection of all functions on $\partial\Omega$ obtained in this way will be denoted by $W^{s-1/p,p}(\partial\Omega)$. For $g \in W^{s-1/p,p}(\partial\Omega)$, we put

$$(0.2.18) \quad [g]_{s-1/p,\partial\Omega} = \inf \|u\|_{W^{s,p}(\Omega)}$$

where the infimum is to be taken over all $u \in W^{s,p}(\Omega)$ satisfying $u|\partial\Omega = g$. Equation (0.2.18) possesses all the properties of the norm, and $W^{s-1/p,p}(\partial\Omega)$ becomes a Banach space with this norm. It is clear that $D^\alpha u|\partial\Omega \in W^{m-|\alpha|-1/p,p}(\partial\Omega)$ if $u \in W^{m,p}(\Omega)$ for some $m \in \mathbb{N}$ and $|\alpha| < m$. We set

$$\tilde{W}^{2-2/p,p}(\Omega) = W^{2-2/p,p}(\Omega), \quad 1 < p \leq \frac{3}{2},$$

$$\tilde{W}^{2-2/p,p}(\Omega) = \{u | u \in W^{2-2/p,p}(\Omega),\ u|\partial\Omega = 0, \text{ where } 0 \text{ is the zero of } W^{2-3/p,p}(\partial\Omega)\}, \quad p > \frac{3}{2}.$$

Let Ω be pathwise connected from now on. According to [FuM] the space $(L^p(\Omega))^n$ can be decomposed as follows:

$$(L^p(\Omega))^n = H_p(\Omega) + \{\nabla g | g \in H^{1,p}(\Omega)\},$$

$+\infty > p > 1$. Here $H_p(\Omega)$ is the completion of the divergence free $(C^\infty_o(\Omega))^n$-vector fields with respect to the norm of $(L^p(\Omega))^n$. The intersection of $H_p(\Omega)$ and $\{\nabla g | g \in H^{1,p}(\Omega)\}$ is the element 0. Let $u \in (L^p(\Omega))^n$. Then there are uniquely determined u_1, u_2 with

$$u = u_1 + u_2,$$
$$u_1 \in H_p(\Omega),$$
$$u_2 \in \{\nabla g \mid g \in H^{1,p}(\Omega)\}.$$

We set $Pu = P_p u = u_1$. Then P_p is a bounded linear operator from $(L^p(\Omega))^n$ onto $H_p(\Omega)$; P_p is called the projection from $(L^p(\Omega))^n$ onto its divergence free part $H_p(\Omega)$. If $p = 2$ the operator P_2 is an orthogonal projection. P_p does not preserve boundary values but regularity; this follows from the way P_p is constructed in [FuM]. In particular this means that

(0.2.19) $P_p((H^{m,p}(\Omega))^n) \subset (H^{m,p}(\Omega))^n,$

(0.2.20) $P_p((C^{k+\alpha}(\bar{\Omega}))^n) \subset (C^{k+\alpha}(\bar{\Omega}))^n,$

$m \in \mathbb{N}$, $k \in \mathbb{N} \cup \{0\}$, $\alpha \in (0,1)$, with continuous imbeddings (see also [Sol1] for (0.2.20)). From (0.2.19) it follows that

(0.2.21) $P_p((H^{s,p}(\Omega))^n) \subset (H^{s,p}(\Omega))^n,\ P_p((W^{s,p}(\Omega))^n) \subset (W^{s,p}(\Omega))^n,$ $s \geq 0,$

with continuous imbeddings.

Let $p = 2$. Then $(H^{-1,2}(\Omega))^n$ has an orthogonal decomposition into $H_{-1,2}(\Omega) = \overline{\{\varphi \mid \varphi \in (C_0^\infty(\Omega))^n, \nabla\cdot\varphi = 0\}}^{\|\cdot\|_{(H^{-1,2}(\Omega))^n}}$ and its orthogonal complement. The projection of $(H^{-1,2}(\Omega))^n$ on $H_{-1,2}(\Omega)$ is denoted by $P_{-1,2}$. Obviously

$$P_{-1,2}(L^2(\Omega))^n = P_2.$$

Moreover we have for p,q, $1 < p,q < +\infty$, $q^+ = \max(p,q)$, $q^- = \min(p,q)$ the relations

(0.2.22) $P_q \mid (L^{q+}(\Omega))^n = P_{q+},$

(0.2.23) $P_p|(L^{q+}(\Omega))^n = P_{q+}$,

(0.2.24) $P_{q-}|(L^p(\Omega))^n = P_p$,

(0.2.25) $P_{q-}|(L^q(\Omega))^n = P_q$.

Finally we mention that $H_p(\Omega) = P_p(L^p(\Omega))^n$ is a reflexive Banach space.

Let $T > 0$, $+\infty > p > 1$. We take the decomposition of $(L^p(\Omega))^n$ just introduced and a vector field $\nabla\pi(t)$ in the second component of this decomposition a.e. on $(0,T)$. In particular this means that a.e. $\pi(t) \in H^{1,p}(\Omega)$; if we assume that $\nabla\pi \in L^p((0,T), (L^p(\Omega))^n)$ then it follows from Bogovskij's result that $\tilde{\pi}(t) = \pi(t) - \frac{1}{\text{mes } \Omega} \int \pi(t)\, dx$ defines a measurable mapping from $(0,T)$ into $L^p(\Omega)$. Poincaré's inequality

$$\|\tilde{\pi}(t)\|_{L^p(\Omega)} \leq c\|\nabla\pi(t)\|_{(L^p(\Omega))^n}$$

shows that $\tilde{\pi} \in L^p((0,T), L^p(\Omega))$. Thus, if $\nabla\pi \in L^p((0,T), (L^p(\Omega))^n)$, we can always assume that $\pi \in L^p((0,T), L^p(\Omega))$.

As usual $\mathbb{N}$ is the set of positive integers, $\mathbb{R}$ is the set of real numbers, $\mathbb{C}$ the set of complex numbers. $\mathbb{R}^+$ is the set of all nonnegative reals, but sometimes we write $[0,\infty)$ or $[0,+\infty)$ instead of it. We will use also the notations $(0,\infty)$, $(0,+\infty)$, $(a,\infty) = (a,+\infty)$, $[a,\infty) = [a,+\infty)$, $[a]$ = largest integer $\leq a$, $(a \in \mathbb{R})$. For Banach spaces X,Y the Banach space of all bounded everywhere defined operators from X into Y is denoted by $L(X,Y)$. For a linear operator $A: D(A) \to Y$ we denote by $D(A) \subset X$ its domain of definition, by $R(A)$ its range, by $N(A)$ its kernel and by $\rho(A)$ its resolvent set $\{z \mid z \in \mathbb{C},\ (z-A)^{-1} \in L(Y,X)\}$. A^* is the adjoint of A.

$c, \tilde{c}$ denote constants, not necessarily the same at each occurence.

I. Linear Equations of Parabolic Type

In this chapter we want to discuss linear equations of parabolic type in a Banach space B, this means equations of the form

$$u' + A(t)u = f,$$
$$u(0) = \varphi,$$

where each $-A(t)$ generates an analytic semigroup in B. As for the domain of definition $D(A(t))$ of $A(t)$ we concentrate on the time-independent case, namely $D(A(t)) = D(A(0))$. The basic material will be put together only briefly since it is common part now of many well known textbooks, e.g. [F], [Y1], [Ta], [Kr].

§ 1. Analytic Semigroups

The notion of a continuous semigroup including the Hille-Yosida Theorem is assumed to be known to the reader, see e.g. [F, Part 2,1].

Let A be a closed operator in a Banach space B with dense domain of definition D(A). A is said to be of type (ϕ, M) if:

(A1) <u>The resolvent set</u> $\rho(-A)$ <u>of</u> $-A$ <u>contains the sector</u> $\Sigma_\phi = \{\lambda \mid \lambda \in \mathbb{C},\ \lambda \neq 0,\ \frac{1}{2}\pi + \phi > \arg \lambda > \frac{3}{2}\pi - \phi\}$ <u>for some</u> $\phi \in (0, \frac{1}{2}\pi)$ <u>and</u>

$$\|(\lambda + A)^{-1}\| \le \frac{M}{|\lambda|},\quad \lambda \in \Sigma_\phi .$$

As it is well known then the following theorem holds:

Theorem I.1.1: $-A$ generates a strongly continuous semigroup e^{-tA} with the following properties

$$(I.1.1)\quad \begin{cases} e^{-tA}x \in D(A),\ x \in B,\ t > 0, \\ Ae^{-tA}x \in C^{o}((0,\infty),B), \end{cases}$$

$$(I.1.2)\quad \begin{cases} \|e^{-tA}\| \leq M,\ t \geq 0, \\ \|Ae^{-tA}x\| \leq \frac{c(M)}{t},\ t > 0. \end{cases}$$

Proof: The proof is well known and rests on the integral representation

$$(I.1.3)\quad e^{-tA} = \frac{1}{2\pi i}\int_{\Gamma} e^{\lambda t}(\lambda+A)^{-1}\, d\lambda,$$

where $\Gamma = \{re^{i\theta_2} | 1 \leq r < +\infty\} \cup \Gamma' \cup \{re^{i\theta_1} | 1 \leq r < +\infty\}$ for some θ_2, θ_1, $\frac{3}{2}\pi - \phi < \theta_2 < \frac{3}{2}\pi$, $\frac{1}{2}\pi + \phi > \theta_1 > \frac{1}{2}\pi$ and $\Gamma' = \{e^{i\theta} | \theta_1 \geq \theta \geq 0 \text{ or } \theta_2 \leq \theta \leq 2\pi\}$; see e.g. [F, Part 2].

In fact e^{-tA} can be continued analytically (with respect to $B(X,X)$) into the sector $\{z | z \neq 0,\ |\arg z| < \phi\}$ and a continuous semigroup e^{-tA} generated by an operator $-A$, A of class (ϕ, M), is called an analytic semigroup. If the condition on $\|(\lambda+A)^{-1}\|$ is replaced by the stronger condition

$$\|(\lambda+A)^{-1}\| \leq \frac{M}{|\lambda| + \omega},\ \lambda \in \Sigma_{\phi},$$

for some $\omega > 0$, then we get the stronger estimates

$$(\text{I.1.4})\quad \begin{cases} \|e^{-tA}\| \leq c(M)e^{-\frac{\omega}{M}t}, & t \geq 0, \\ \|Ae^{-tA}\| \leq c(M)\dfrac{e^{-\frac{\omega}{M}t}}{t}, & t > 0. \end{cases}$$

These estimates follow from the integral representation (I.1.3):

We check the first one. The resolvent series

$$(\mu+A)^{-1} = \sum_{n=0}^{\infty} (\lambda-\mu)^n(\lambda+A)^{-(n+1)}, \quad \lambda \in \Sigma_\phi,$$

shows that all μ, $|\mu-\lambda| < \frac{1}{M}(|\lambda|+\omega)$, are also in $\rho(A)$, thus all $\mu = \lambda-\delta$ with $\mathbb{R}^+ \ni \delta < \frac{\omega}{M}$. Then $(\mu+A)^{-1}$ satisfies the estimate

$$\|(\mu+A)^{-1}\| \leq \sum_{n=0}^{\infty} \frac{M^{n+1}\delta^n}{(|\lambda|+\omega)^{n+1}},$$

$$\leq \frac{M}{|\lambda|+\omega-M\delta},$$

$$\leq \frac{M}{|\lambda|}.$$

Thus $A-\delta$ is of type (ϕ,M). Since $-(A-\delta)$ generates $e^{\delta t}e^{-At}$ we arrive at the desired result, if we let δ tend to ω/M. □

If $m \in \mathbb{N}$ we get from the semigroup property of e^{-tA} the well known result that

$$e^{-tA}x \in D(A^m), \quad t > 0, \quad x \in B,$$

$$A^m e^{-tA}x \in C^0((0,\infty),B),$$

$$\|A^m e^{-tA}\| \leq \frac{c(M,m)}{t^m}.$$

§ 2. The Evolution Operator U(t,s)

Let A be of type (ϕ,M). Then $e^{-tA}\varphi$, $\varphi \in D(A)$, is the unique solution of $u' + Au = 0$, $u(0) = \varphi$ over any interval $[0,T]$ within the class $u \in C^1([0,T],B)$, $u(t) \in D(A)$, $0 \le t \le T$, $Au \in C^0([0,T],B)$, $u(0) = \varphi$. The same question will now be studied for $u' + A(t)u = 0$. We need the following assumptions:

(U1) The A(t) are closed operators in the Banach space B with constant domain of definition $D(A(t)) = D(A(0))$, $0 \le t \le T$.

(U2) $\{\lambda \mid \mathrm{Re}\ \lambda \ge 0\} \subset \rho(-A(t))$ and

$$||(\lambda+A(t))^{-1}|| \le \frac{M'}{|\lambda|+1}, \quad t \in [0,T],$$

(U3) $||(A(t)-A(r))A^{-1}(s)|| \le \hat{c}|t-r|^{\alpha}$, $0 \le t,r,s \le T$, for some $\alpha \in (0,1]$. $\hat{c}$ is a positive constant to be used later on.

Taking the resolvent series

$$(\mu+A(t))^{-1} = \sum_{n=0}^{\infty} (\lambda-\mu)^n (\lambda+A(t))^{-(n+1)}$$

we see that all μ with $\mathrm{Im}\ \mu = \mathrm{Im}\ \lambda$, $|\mathrm{Re}\ \mu| \le |\mathrm{Im}\ \lambda| \frac{1}{(1+\varepsilon)M'}$ are in $\rho(A(t))$ if λ is given with $\mathrm{Re}\ \lambda = 0$, $\varepsilon > 0$. Moreover

(I.2.1) $$||(\mu+A(t))^{-1}|| \le \frac{M'}{|\lambda|+1} \frac{1}{1-\frac{M'}{|\lambda|+1}|\lambda-\mu|} \le \frac{M'(1+\varepsilon)/\varepsilon}{|\lambda|+1},$$

$$\le \frac{\sqrt{1+(\frac{1}{M'(1+\varepsilon)})^2}\, M'(1+\varepsilon)/\varepsilon}{|\mu| + \sqrt{1+(\frac{1}{M'(1+\varepsilon)})^2}}.$$

Thus each A(t) is of type $(\phi_\varepsilon, \frac{1+\varepsilon}{\varepsilon}M'\sqrt{1+(\frac{1}{M'(1+\varepsilon)})^2})$ with $\sin \phi_\varepsilon \le$ $\le \frac{1}{(1+\varepsilon)M'}/(1+(\frac{1}{(1+\varepsilon)M'})^2)^{1/2}$, $\varepsilon > 0$. $-A(t)$ therefore generates an analytic semigroup which decays exponentially with any exponent $\delta < 1/M'$.

Theorem I.2.1: There exists a uniquely determined operator valued function U: $\{(t,s) \mid (t,s) \in [0,T]\times[0,T],\ s \le t\}$ with the following properties: U is strongly continuous, U is strongly continuously differentiable on $t > s$ with respect to t, $U(t,s)D(A(0)) \subset D(A(0))$,

$$\text{(I.2.1)}\quad \begin{cases} \dfrac{\partial U(t,s)x}{\partial t} + A(t)U(t,s)x = 0, \quad U(t,t) = I, \\ \qquad\qquad U(t,s)U(s,r) = U(t,r). \end{cases}$$

U(t,s) is called the evolution operator generated by the -A(t).

The proof can be found in [F, Part 2] or [Ta, 5.2]. It is based on the integral representation

$$U(t,s) = e^{-(t-s)A(s)} + \int_s^t e^{-(t-\sigma)A(\sigma)}\phi(\sigma,s)\, d\sigma,$$

where

$$\phi(\sigma,s) = \phi_1(\sigma,s) + \int_s^\sigma \phi(\sigma,\tau)\phi_1(\tau,s)\, ds,$$

$$\phi_1(\sigma,s) = (A(s)-A(\sigma))e^{-(\sigma-s)A(s)}.$$

Thus ϕ could be expanded in a series

$$\phi(\sigma,s) = \sum_{k=1}^{\infty} \phi_k(\sigma,s)$$

with

$$\phi_{k+1}(\sigma,s) = \int_s^\sigma \phi_k(\sigma,\tau)\phi_1(\tau,s)\, d\tau.$$

Now we want to estimate $\|U(t,s)\|$. Setting $M'_\varepsilon = \frac{1+\varepsilon}{\varepsilon}\sqrt{1+(\frac{1}{M'(1+\varepsilon)})^2}M'$ and using (U3) we see by induction that for $\varepsilon = \varepsilon(\delta)$

$$\|\phi_k(\sigma,s)\| \le \frac{(\hat{c}\Gamma(\alpha)c(M'_\varepsilon))^k e^{-\delta(\sigma-s)}|\sigma-s|^{k\alpha}}{\Gamma(k\alpha)|\sigma-s|}, \quad \delta < 1/M'.$$

Thus the series for ϕ converges uniformly on $T \ge \sigma > s \ge 0$ in the topology of L(B,B) and we have

$$\|\phi(\sigma,s)\| \le \sum_{k=1}^{\infty} \frac{(\hat{c}\Gamma(\alpha)c(M'_\varepsilon))^k |\sigma-s|^{k\alpha}}{\Gamma(k\alpha)} \cdot \frac{e^{-\delta(\sigma-s)}}{|\sigma-s|},$$

$$\le \hat{c}\Gamma(\alpha)c(M'_\varepsilon)\frac{e^{-\delta(\sigma-s)}}{|\sigma-s|^{1-\alpha}} \sum_{k=0}^{\infty} \frac{(\hat{c}\Gamma(\alpha)c(M'_\varepsilon))^k |\sigma-s|^{k\alpha}\Gamma(k\alpha+1)}{\Gamma(k\alpha+1)\Gamma(k\alpha+\alpha)},$$

$$\le \hat{c}\Gamma(\alpha)c(M'_\varepsilon)c(\eta,\alpha)\frac{e^{-\delta(\sigma-s)}}{|\sigma-s|^{1-\alpha}} \cdot e^{(\hat{c}\Gamma(\alpha)c(M'_\varepsilon)\eta)^{\frac{1}{\alpha}}(\sigma-s)},$$

$\eta > 1$. Here we substituted η^k/η^k and used Hölder's inequality for series with exponents $1/(1-\alpha)$ and $1/\alpha$ and the inequality $\Gamma(k\alpha+1) \ge \sqrt{2\pi}(k\alpha)^{k\alpha+\frac{1}{2}}e^{-k\alpha}$, whereas $\hat{c}$ stems from (U3). This gives the estimate

$$\text{(I.2.2)} \quad \|U(t,s)\| \le c(\hat{c},\eta,\alpha,M'_\varepsilon)e^{((\hat{c}\Gamma(\alpha)c(M'_\varepsilon)\eta)^{\frac{1}{\alpha}}-\delta)(t-s)}.$$

Choosing ε sufficiently large we see that if $(\hat{c}\Gamma(\alpha)c(M'))^{\frac{1}{\alpha}} < 1/M'$, $\|U(t,s)\|$ even decays exponentially.

More generally the estimates

$$\|U(t,s)\| \le c(T,\alpha),$$

$$\|A(t)U(t,s)\| \le \frac{c(T,\alpha)}{t-s},$$

$$\|A(t)U(t,s)A^{-1}(s)\| \le c(T,\alpha)$$

are well known [F, Part 2], where $c(T,\alpha)$ continuously depends on T,α. It also follows that $U(t,s)x$, $x \in D(A(0))$, is continuously differentiable with respect to s on $t > s$ and

$$\text{(I.2.3)} \quad \frac{\partial U(t,s)x}{\partial s} - U(t,s)A(s)x = 0.$$

Moreover $A(t)U(t,s)A^{-1}(s)$ is strongly continuous in $t \ge s$.

Now it is possible to solve Cauchy's problem for $u' + A(t)u = f(t)$. We have

Theorem I.2.2: *Let the assumptions* (U1)-(U3) *be satisfied. Let* $\varphi \in B$, *let* $f \in C^{\beta}([0,T],B)$ *for some* $\beta \in (0,1]$. *Then there is a unique mapping* $u:[0,T] \to B$ *with*

$$u \in C^{0}([0,T],B) \cap C^{1}((0,T],B)$$

$$u(t) \in D(A(0)),\ 0 < t \leq T,\ A(.)u(.) \in C^{0}((0,T],B),$$

$$u' + A(t)u = f,\ u(0) = \varphi.$$

If $\varphi \in D(A(0))$ *then* $u \in C^{1}([0,T],B)$, $A(.)u(.) \in C^{0}([0,T],B)$; u *is given by*

$$\text{(I.2.4)}\quad u(t) = U(t,0)\varphi + \int_0^t U(t,s)\, f(s)\, ds$$

in any case.

Proof: We give the proof only if instead of (U3) the following assumption holds: $A(.)x \in C^{1}([0,T],B)$, $x \in D(A(0))$. Then

$$||(A(t)-A(r))A^{-1}(0)x|| \leq \int_r^t ||A'(\sigma)A^{-1}(0)x||\, d\sigma,$$

where $A'(t)x := \frac{d}{dt}(A(t)x)$, $x \in A(0)$. By the uniform boundedness principle the norms $||\frac{A(t)-A(r)}{t-r}A^{-1}(0)||$ are uniformly bounded, $t,r \in [0,T]$, $t \neq r$. If follows that $A(t)A^{-1}(0)$ is continuous in the topology of $L(B,B)$ on $[0,T]$ and therefore $A(0)A^{-1}(t)$ has the same property. Thus (U3) holds with $\alpha = 1$.

We have to evaluate the right hand side of (I.2.4). First of all it is clear that $u \in C^{0}([0,T],B)$. We have by (I.2.3)

$$\text{(I.2.5)}\quad \int_0^t U(t,s)f(s)\,ds = \int_0^t U(t,s)(f(s)-f(t))\,ds + \int_0^t U(t,s)f(t)\,ds$$

$$= \int_0^t U(t,s)(f(s)-f(t))\,ds + $$

$$+ \int_0^t \frac{\partial}{\partial s}(U(t,s))A^{-1}(s)f(t)\,ds,$$

$$\text{(I.2.6)}\quad = \int_0^t U(t,s)(f(s)-f(t))\,ds +$$

$$+ [U(t,s)A^{-1}(s)f(t)]_0^t +$$

$$+ \int_0^t U(t,s)A^{-1}(s)A'(s)A^{-1}(s)f(t)\,ds.$$

In virtue of $f \in C^\beta([0,T],B)$ we see that $\int_0^t U(t,s)f(s)ds \in D(A(0))$ moreover $A(t)[U(t,s)A^{-1}(s)f(t)]_0^t + \int_0^t U(t,s)A^{-1}(s)A'(s)A^{-1}(s)f(t)$ ds is continuous. We obtain $(h \geq 0)$

$$\text{(I.2.7)}\quad \int_0^{t+h} A(t+h)U(t+h,s)(f(s)-f(t+h)ds - \int_0^t A(t)U(t,s)(f(s)-f(t))d$$

$$= \int_0^t [A(t+h)U(t+h,s)(f(s)-f(t+h)) - A(t)U(t,s)(f(s)-f(t))]\,d$$

$$+ \int_t^{t+h} A(t+h)U(t+h,s)(f(s)-f(t+h))\,ds.$$

The second integral fulfils the estimate

$$\text{(I.2.8)}\quad \left\|\int_t^{t+h} A(t+h)U(t+h,s)(f(s)-f(t+h))\,ds\right\| \leq c(T,\alpha)\left|\int_t^{t+h} \frac{ds}{(t+h-s)^{1-}}\right.$$

$$\leq c(T,\alpha,\beta)|h|^\beta.$$

The first one tends to 0 for $h \to 0$ by the strong continuity of $A(t)U(t,s)$ on $t > s$, our assumption on f and the Lebesgue-convergence theorem. Thus $A(t)\int_0^t U(t,s)f(s)\,ds$ is continuous, $\int_0^t U(t,s)f(s)\,ds$ is continuously differentiable, and we have

$$\frac{d}{dt}\int_0^t U(t,s)f(s)\,ds = f(t) - A(t)\int_0^t U(t,s)\,f(s)\,ds.$$

Thus our theorem is proved if: $A(.)x \in C^1([0,T],B)$.
If only (U3) is fulfilled then one has to rewrite the second integral in (I.2.5) in the form

$$\int_0^t U(t,s)f(t)\,dt = \int_0^t (U(t,s)-e^{-(t-s)A(t)})f(t)\,dt + $$
$$+ (I-e^{-tA(t)})A^{-1}(t)f(t)$$

and use the estimates

$$\text{(I.2.9)} \quad \|U(t,s)-e^{-(t-s)A(t)}\| \le C(T,\alpha), \quad t,s \in [0,T]$$

$$\text{(I.2.10)} \quad \|A(t)(U(t,s)-e^{-(t-s)A(t)})\| \le C(T,\alpha)\frac{1}{|t-s|^{1-\alpha}}, \quad t,s \in [0,T], \quad t \neq s,$$

see [F, Part 2,7].

§ 3. Fractional Powers

In this section we want to introduce the concept of the fractional power of a closed operator A, where $-A$ generates an analytic semigroup e^{-tA}. References are [F], [Ta] or [Y1].

Let A be a closed operator in a Banach space B with domain of definition D(A). Let $\{\lambda \mid \operatorname{Re}\lambda \ge 0\}$ be in $\rho(-A)$ and

$$\|(\lambda+A)^{-1}\| \le \frac{M'}{|\lambda|+1}, \quad \operatorname{Re}\lambda \ge 0.$$

Thus A fulfils the assumption (U2) of § 2, and -A generates an analytic semigroup e^{-tA} with

$$\|e^{-tA}\| \le c(M'_\varepsilon,\delta)e^{-\delta t}, \quad \delta < 1/M', \quad \varepsilon = \varepsilon(\delta).$$

We define

$$A^{-\alpha} = \frac{1}{\Gamma(\alpha)} \int_0^\infty e^{-sA} s^{\alpha-1}\, ds, \quad \mathrm{Re}\ \alpha > 0.$$

Then $A^{-\alpha} \in L(B,B)$ and $\|A^{-\alpha}\| \le \dfrac{c(M'_\varepsilon,\delta)}{|\Gamma(\alpha)\mathrm{Re}\ \alpha|}$. We get a formula for A^{-1}, which is well known from the theory of continuous semigroups, if $\alpha = 1$.

Now we put together some well known facts on fractional powers and evolution operators which e.g. can be found in [F, Part 2, 14]. We set

$$A^\alpha = (A^{-\alpha})^{-1}$$

which is possible since from $A^{-\alpha}v = 0$ it follows $v = 0$. A^α then is a closed operator with dense domain of definition $D(A^\alpha)$ and

$$D(A^\alpha) \subset D(A^\beta) \text{ if } \mathrm{Re}\ \alpha > \mathrm{Re}\ \beta,$$

$$D(A^\alpha) \subset D(A^\beta) \text{ if } \alpha \ge \beta,\ \alpha,\beta \in \mathbb{R},$$

(I.3.1) $A^\alpha A^\beta v = A^\beta A^\alpha v = A^{\alpha+\beta} v,$

$v \in D(A^\gamma)$, $\gamma = \max(\mathrm{Re}\ \alpha, \mathrm{Re}\ \beta, \mathrm{Re}(\alpha+\beta))$ for $\mathrm{Re}(\alpha+\beta) \neq 0$ and $\gamma = \max(\mathrm{Re}\ \alpha, \mathrm{Re}\ \beta) > 0$ if $\mathrm{Re}(\alpha+\beta) = \alpha+\beta = 0$. If $\alpha,\beta \in \mathbb{R}$ then (I.3.1) holds for $v \in D(A^\gamma)$, $\gamma = \max(\alpha,\beta,\alpha+\beta)$.

Now let operators $A(t)$, $0 \le t \le T$, be given as in § 2. We want to study the operators $A^\zeta(t)e^{-(t-s)A(t)}$ and $A^\zeta(t)U(t,s)$, $\mathrm{Re}\ \zeta > 0$. From the representation of $A^{-\zeta}(t)$ it follows that $A^{-\zeta}(t)$ is a holomorphic mapping from

$$\{\zeta \mid \zeta \in \mathbb{C},\ \mathrm{Re}\ \zeta > 0\} \text{ into } L(B,B).$$

If $0 < \mathrm{Re}\ \zeta < 1$ we have

$$A^{\zeta}(t)e^{-A(t)\tau} = \frac{1}{2\pi i}A^{\zeta-1}(t)\int_{\Gamma} e^{\lambda\tau}A(t)(\lambda+A(t))^{-1}\,d\lambda,\quad \tau > 0,$$

and

$$\text{(I.3.2)}\quad \|A^{\zeta}(t)e^{-A(t)\tau}\| \leq \frac{c(M')}{(1-\mathrm{Re}\ \zeta)|\Gamma(1-\zeta)|\tau}.$$

Thus $A^{\zeta}(t)e^{-A(t)\tau}$ depends holomorphically on ζ, $1 > \mathrm{Re}\ \zeta > 0$, for each $\tau > 0$. Now let $\mathrm{Re}\ \zeta > 0$, $|\zeta| < M \in \mathbb{N}$. Then $0 < \mathrm{Re}\ \frac{\zeta}{M} < 1$ and

$$A^{\zeta}(t)e^{-A(t)\tau} = \prod_{1}^{M} A^{\frac{\zeta}{M}}(t)e^{-A(t)\frac{\tau}{M}}.$$

Therefore $A^{\zeta}(t)e^{-A(t)\tau}$ is holomorphic with respect to the $L(B,B)$-topology in $\mathrm{Re}\ \zeta > 0$. Moreover

$$A^{\zeta}(t)e^{-A(t)\tau}u = e^{-A(t)\tau}A^{\zeta}(t)u,$$

$u \in D(A^{\zeta}(t))$, $\mathrm{Re}\ \zeta \neq 0$.

Actually we will need estimates for $A^{\zeta}(t)U(t,s)$ only in the case that ζ is real. Thus we now concentrate on this case. As it is proved in [F, Part 2, 14] we then have the following estimates

$$\text{(I.3.3)}\quad \|A^{\beta}(t)v\| \leq C(\alpha,\beta,\gamma)\|A^{\gamma}(t)v\|^{(\beta-\tilde{\alpha})/(\gamma-\tilde{\alpha})}\|A^{\tilde{\alpha}}(t)v\|^{(\gamma-\beta)/(\gamma-\tilde{\alpha})},$$

$$v \in D(A^{\gamma}(t)),\ \tilde{\alpha} < \beta < \gamma,$$

$$\text{(I.3.4)}\quad \|A^{\beta}(t)e^{-A(t)\tau}\| \leq \frac{C(M'_{\varepsilon},\beta)}{\tau^{\beta}}e^{-\delta\tau},\quad 0 \leq t \leq T,\ \delta < 1/M',\ \varepsilon = \varepsilon(\delta),$$

$$\text{(I.3.5)}\quad \|A^{\gamma}(t)[A^{-\beta}(t)-A^{-\beta}(\tau)]\| \leq C(\gamma,\beta,\alpha)|t-\tau|^{\alpha},\quad 0 \leq \gamma < \beta < 1,$$

$$0 \leq \tau,t \leq T,$$

(I.3.6) $\|A^{\gamma}(t)e^{-(t-\tau)A(t)}A^{-\beta}(\tau)\| \leq C(M'_{\varepsilon},\gamma,\beta,\alpha)(1+|t-\tau|^{\alpha})e^{-\delta(t-\tau)}\cdot |t-\tau|^{\beta-\tilde{\varepsilon}-\gamma}$,

$0<\beta<1,\ 0<\tilde{\varepsilon}<\beta,\ \beta\leq\gamma,\ 0\leq\tau<t\leq T,\ \delta<1/M',\ \varepsilon=\varepsilon(\delta)$,

(I.3.7) $\|A^{\gamma}(t)U(t,\tau)A^{-\beta}(\tau)\| \leq C(M',\gamma,\beta,\alpha,T)\,|t-\tau|^{\beta-\tilde{\varepsilon}-\gamma}$,

$0<\beta<1,\ 0<\tilde{\varepsilon}<\beta,\ \beta\leq\gamma<1+\alpha,\ 0\leq\tau<t\leq T$.

$A^{\gamma}(t)U(t,\tau)A^{-\beta}(\tau)x$ is continuous in $t>\tau$, $x\in B$.

If $A(t)x$, $x\in D(A(0))$, is continuously differentiable, then $A^{\beta}(t)x$ $0\leq\beta<1$, is too for every x being contained in a $D(A^{\beta+\varepsilon}(\tau))$ for any $\varepsilon\in(0,1-\beta]$, $\tau\in[0,T]$, and

(I.3.8) $\|A^{\beta'}(t)x\| \leq C(M',\varepsilon,\beta,T)\|A^{\beta+\varepsilon}(\tau)x\|$.

For a proof see [Kr, p. 181-184].

§ 4. Comments to Chapter I

<u>To § 1 and § 2:</u> Let $B=L^{p}(\Omega)$ with $\infty>p>1$ and a bounded open set $\Omega\subset\mathbb{R}^{n}$. The boundary of Ω is assumed to be sufficiently smooth, say of class C^{∞}. Moreover Ω is locally on one side of Ω. Let $m\in\mathbb{N}$, $T>0$. For every multiindex α of $\mathbb{R}^{n}$ with $|\alpha|\leq 2m$ let there functions $A_{\alpha}:[0,T]\times\bar{\Omega}\to\mathbb{R}$ for $|\alpha|=2m$, $A_{\alpha}:(0,T)\times\Omega\to\mathbb{C}$, $|\alpha|\leq 2m-1$ be given with the following properties:

(I.4.1) $A_{\alpha}\in C^{0}([0,T],C^{0}(\bar{\Omega}))$, $|\alpha|=2m$,

(I.4.2) $A_{\alpha}\in L^{\infty}((0,T)\times\Omega)$, $|\alpha|\leq 2m-1$,

(I.4.3) $\underset{x\in\Omega}{\operatorname{ess\,sup}}\,|A_{\alpha}(t,x)-A_{\alpha}(r,x)|\leq c|t-r|^{\tilde{\alpha}}$, $t,r\in[0,T]$,

for some $\tilde{\alpha}\in(0,1)$, $|\alpha|\leq 2m$,

(I.4.4) $\sum_{|\alpha|=2m} A_\alpha(t,x)\xi^\alpha \geq c_o|\xi|^{2m}$, $t\in[0,T]$, $x\in\bar{\Omega}$, $\xi\in\mathbb{R}^n$

for some $c_o>0$. The latter condition is called condition of ellipticity. In the terminology of [Br] the operators

$$A(t) = \sum_{|\alpha|\leq 2m} A_\alpha(t,x)D^\alpha$$

are essentially real and uniformly strongly elliptic. We consider the A(t) under Dirichlet boundary conditions, i.e.

$$A(t)u = \sum_{|\alpha|\leq 2m} A_\alpha(t)D^\alpha u, \quad u\in D(A(t)) = D(A(0)) = H^{2m,p}(\Omega)\cap \cap \overset{o}{H}{}^{m,p}(\Omega).$$

Then the A(t) are closed operators in $L^p(\Omega)$ with dense domain of definition since

(I.4.5) $\|u\|_{2m,p} \leq c(\|A(t)u\|_{o,p} + \|u\|_{o,p})$, $u\in D(A(0))$,

where $c = c(m,n,\Omega,c_o,\omega_x,\|A_\alpha\|_{L^\infty((0,T)\times\Omega)})$. ω_x is the modulus of continuity in x for the leading coefficients, i.e.

$$\omega(r) = \sup_{\substack{|\alpha|=2m,\\ 0\leq t\leq T,\\ |x-y|\leq r}} |A_\alpha(t,x) - A_\alpha(t,y)|.$$

If $Au = 0$ in $H^{2m,p}(\Omega)\cap \overset{o}{H}{}^{m,p}(\Omega)$ only admits the solution $u=0$ then (I.4.5) can be strengthened, i.e. we have

(I.4.6) $\|u\|_{2m,p} \leq c\|A(t)u\|_{o,p}$,

references for these results are [Br ,p. 44], Agmon, Douglis and Nirenberg [ADN I, p. 704]. Agmon [Ag] has proved that all λ with $\mathrm{Re}\,\lambda \geq \Lambda_o$ for some $\Lambda_o > 0$ are in the resolvent set of A(t), $0\leq t\leq T$, and that

$$\text{(I.4.7)}\quad \sum_{j=0}^{2m} |\lambda|^{(2m-j)/m}\|u\|_{j,p} \leq c\|(A(t)+\lambda)u\|_{o,p}, \ \operatorname{Re}\lambda \geq \Lambda_o,$$

where c, Λ_o depend on the same quantities as c above. Thus by the resolvent series we get

$$\text{(I.4.8)}\quad \|(A(t)+\Lambda_o+\lambda)^{-1}\| \leq \frac{M'}{|\lambda|+1}, \ \operatorname{Re}\lambda \geq 0.$$

If all λ, $\operatorname{Re}\lambda \geq 0$, are in the resolvent set of $A(t)$, then (I.4.7) holds for all λ with $\operatorname{Re}\lambda \geq 0$, and we get

$$\text{(I.4.9)}\quad \|(A(t)+\lambda)^{-1}\| \leq \frac{M'}{|\lambda|+1}, \ \operatorname{Re}\lambda \geq 0.$$

This also follows by use of the resolvent series.

So far we have not used (I.4.3). This assumption together with the previous estimates shows that the $A(t)+\Lambda_o$ fulfil (U1)-(U3). Thus we can construct $U(t,s)$ and solve the parabolic initial boundary value problem $u' + A(t)u = f$, $u(0) = \varphi$, $u(t) \in \overset{o}{H}{}^{m,p}(\Omega)$, $0 < t \leq T$: $U(t,s)$ is the evolution operator to $A(t)+\Lambda_o$, and $u(t) = e^{\Lambda_o t}U(t,0)$ is the desired solution in the homogeneous case.

As it is seen from § 1 the question if e^{-tA} decays exponentially essentially depends from the upper bound of $\sigma(-A)$. Such a result is not known for evolution operators $U(t,s)$, even in the special case of the elliptic operators $A(t)$ just introduced. In fact, we only have given a perturbative result in § 2 which requires that the $A(t)$ do not oscillate too strongly.* In virtue of the maximum principle the situation is different for $m = 1$. Let the following additional assumptions be fulfilled: $m = 1$, $A_\alpha \in C^o([0,T], C^{1/2}(\bar{\Omega}))$, $i^{|\alpha|} \cdot A_\alpha(t,x) \in \mathbb{R}$, $0 \leq t \leq T$, $x \in \bar{\Omega}$, $|\alpha| \leq 2$; let $A_\alpha(t,x) \geq 0$ on $[0,T]\times\bar{\Omega}$, $|\alpha| = 0$. Let $\tilde{\alpha} = \frac{1}{2}$. Let $\tilde{A}(t) = A(t)+\Lambda_o$,

$$\delta = \frac{M^2}{4c_o^2 n}, \text{ where } \|A_\alpha(t)\|_{C^o(\overline{\Omega})} \leq M, \ 1 \leq |\alpha| \leq 2,$$

$$\lambda = e^{\delta R^2}+1, \text{ where } R^2 \geq |x|^2, \ x \in \overline{\Omega},$$

$$\gamma = \frac{\delta n c_o}{2} e^{-\delta R^2}.$$

Let u be a real solution of

$$u' + A(t)u = 0, \text{ i.e. } u' + (A(t)+\Lambda_o)u = \Lambda_o u,$$
$$u(0) = \varphi \in D(A(0)), \quad \varphi \text{ real},$$
$$u(t) \in \overset{o}{H}{}^{1,p}(\Omega).$$

Choosing $p > n$ we see by Sobolev's imbedding theorem that $u \in C^o([0,T],C^{1+\varepsilon}(\overline{\Omega}))$ for some $\varepsilon > 0$. Since

$$u(t) = e^{\Lambda_o t} U(t,0)\varphi$$

we see by (I.3.7) that $u(t) \in D(\widetilde{A}^\gamma(t))$, $0 < t \leq T$, $\widetilde{A}^\gamma(.)u(.) \in$ $\in C^o((0,T],L^p(\Omega))$, $\gamma < 1+\widetilde{\alpha}$. Let us choose $\gamma \in (\frac{n}{2p}+1, 1+\widetilde{\alpha})$. Then $D(\widetilde{A}^{\gamma-1}(t)) \subset C^\varepsilon(\overline{\Omega})$ and $\|v\|_{C^\varepsilon(\overline{\Omega})} \leq c\|\widetilde{A}^{\gamma-1}(t)v\|$, $0 \leq t \leq T$, for some $\varepsilon \in (0,\frac{1}{2})$. This follows from Sobolev's imbedding theorem in 0.2 and [F, pp. 177,178]. Since $\widetilde{A}^{\gamma-1}(t)\widetilde{A}^{-(\gamma-1+\varepsilon')}(\tau)$ is even continuous in the L(B,B)-norm with respect to τ in t (see I.3.5) and since $\widetilde{A}^{\gamma-1+\varepsilon'}(\tau)\widetilde{A}(\tau)U(\tau,0)\varphi$ is continuous on $(0,T]$ for some $\varepsilon' > 0$ we see that $\widetilde{A}(t)U(t,0)\varphi$ is continuous in $C^\varepsilon(\overline{\Omega})$ in $(0,T]$. By the elliptic regularity theory (see [ADN1, Theorem 7.3]) we get that $u(t) \in C^o((0,T],C^{2+\varepsilon}(\overline{\Omega}))$. It follows $u'(t) \in C^o((0,T],C^\varepsilon(\overline{\Omega}))$. Since $p > n$ and since the continuity of $\widetilde{A}(.)u(.)$ means that $u(.) \in$ $\in C^o([0,T],H^{2,p}(\Omega))$ we see that in particular $u \in C^o([0,T]\times\overline{\Omega})$.

On u can be applied the classical maximum principle for parabolic equations. This we will do in the following way: Let

$$\omega(t,x) = e^{-\gamma t}(\lambda - e^{\delta|x|^2})\|\varphi\|_{C^o(\overline{\Omega})}.$$

Then an easy calculation shows that

$$\omega'(t,x) + \sum_{|\alpha|\leq 2} A_\alpha(t,x)D^\alpha\omega(t,x) \geq 0,$$
$$\omega(t,x)\,|\,\partial\Omega \geq 0,$$
$$\omega(0,x) \geq \|\varphi\|_{C^0(\bar{\Omega})}.$$

Thus for u and -u we get that in (t,x) they are $\leq \omega(t,x)$. Therefore $\|u(t)\|_{C^0(\bar{\Omega})}$ decays exponentially. *

To § 3: As it is seen from (I.3.5) we do not know if $D(A^\gamma(t)) = D(A^\gamma(0))$, $0 \leq t \leq T$, $0 < \gamma < 1$, although we assumed that $D(A(t)) = D(A(0))$. The situation is different if B is a Hilbert space H and the A(t) are selfadjoint with $(A(t)u,u) \geq c\|u\|^2$, $u \in D(A(0))$, for some $c > 0$. Then Heinz [H1] proved that $D(A^\gamma(t)) = D(A^\gamma(0))$, $\|A^\gamma(t)x\| \leq c\|A^\gamma(0)x\|$, $0 \leq t \leq T$, $0 \leq \gamma \leq 1$, $x \in D(A^\gamma(0))$. In this case we also have

(I.4.10) $A^\gamma(.)x \in C^1([0,T],H)$,

(I.4.11) $\|A^{\gamma'}(t)x\| \leq c\|A^\gamma(0)\|$, $x \in D(A^\gamma(0))$,

if $A(.)x \in C^1([0,T],H)$, $x \in D(A(0))$ (see [Kr, II.1]). (I.4.10,11) are a consequence of the result by Heinz just mentioned above and the techniques of the proof employed by Heinz.

* Let us remain in a Hilbert space H. If A(t) is maximal accretive and moreover

$$\mathrm{Re}(A(t)u,u) \geq c\|u\|^2, \quad u \in D(A(0)),$$

for some $c > 0$, Kato [K3, K4] then proved that

(I.4.12) $D(A^{\gamma}(t)) = D(A^{\gamma}(0))$,

(I.4.13) $\|A^{\gamma}(t)x\| \leq c\|A^{\gamma}(0)x\|$, $0 \leq t \leq T$, $0 \leq \gamma \leq 1$, $x \in D(A^{\gamma}(0))$.

Assertions like (I.4.10)-(I.4.13) are important in applications (see e.g. [K3 , K4], [W1]). *

II. Local Solutions of First Order Semilinear Evolution Equations

§ 1. Solutions of Equations with Nonlinearities Relatively Bounded to A

Let B be a reflexive Banach space, let A be a closed operator in B with dense domain of definition D(A). Let all λ, Re $\lambda \geq 0$, be in the resolvent set of -A and let $\|(\lambda+A)^{-1}\| \leq \frac{M'}{|\lambda|+1}$ for all λ, Re $\lambda \geq 0$. This means that -A generates an analytic semigroup e^{-tA} (c (I.2)). We consider here the nonlinear initial value problem

$$\text{(II.1.1)} \quad \begin{cases} u' + Au + M(u) = 0, \\ \qquad\qquad u(0) = \varphi \in D(A) \end{cases}$$

in B. Usually then M is a locally Hölder [1] or Lipschitz continuous mapping from $D(A^{1-\rho})$ into B for some ρ, $0 < \rho < 1$. $A^{1-\rho}e^{-tA}x$ is continuous for $t > 0$, $x \in B$, $1 \geq \rho \geq 0$ and fulfills the estimate $\|A^{1-\rho}e^{-tA}x\| \leq c(\delta,M')e^{-\delta t}/t^{1-\rho}$, $\delta < 1/M'$, $0 \leq t \leq T$. Under the assumption above on M one can show that (II.1.1) has a unique local (in time) strong solution on a maximal interval $[0,T(\varphi))$ with a $T(\varphi)$, $0 < T(\varphi) \leq \infty$. In this paragraph we suppose first that M is a Lipschitz continuous mapping from D(A) into B only, satisfying the following Lipschitz condition:

$$\text{(II.1.2)} \quad \begin{cases} \|M(u)-M(v)\| \leq k(C)\|A^{1-\rho}(u-v)\|, \ \|M(u)\| \leq k(C), \\ u,v \in D(A), \ \|Au\|+\|Av\| \leq C \end{cases}$$

for some ρ, $0 < \rho < 1$.

---

[1] If A^{-1} is compact. For an oversight over various conditions on M and their consequences see "Comments to chapter II, § 1".

If $\{\lambda | \operatorname{Re}\lambda \geq 0\}$ is in the resolvent set of $-A$ but $-A$ generates a continuous semigroup e^{-tA} only we can show the existence of a unique local (in time) strong solution of (II.1.1) if M is a Lipschitz continuous mapping from D(A) into B too but fulfils the stronger Lipschitz condition

$$\text{(II.1.3)} \quad \begin{cases} \|M(u)-M(v)\| \leq k(C)\|u-v\|, \ \|M(u)\| \leq k(C), \\ u,v \in D(A), \ \|Au\|+\|Av\| \leq C. \end{cases}$$

This case is treated here too for the sake of completeness.

First we need a differentiability lemma on M.

<u>Lemma II.1.1:</u> 1. <u>Let</u> M <u>fulfill the Lipschitz condition</u> (II.1.3). <u>Let</u> $T>0$, <u>let</u> $w \in C^1([0,T],B)$, $w(t) \in D(A)$, $0 \leq t \leq T$, $Aw \in C^0([0,T],B)$. <u>Then</u> M(w) <u>is weakly differentiable with derivative</u> M(w)', <u>i.e.</u> <u>in particular</u> $\int_0^T \langle M(w)'(s),\psi(s)\rangle\, ds = -\int_0^T \langle M(w(s)),\psi'(s)\rangle\, ds$ <u>for</u> <u>all</u> $\psi \in C^1([0,T],B^*)$ <u>with compact support in</u> $(0,T)$. <u>Moreover</u> $\sup_{0\leq s\leq T} \|M(w)'(s)\| \leq k(2 \sup_{0\leq s\leq T} \|Aw(s)\|) \sup_{0\leq s\leq T} \|w'(s)\|$.

2. <u>Let</u> M <u>fulfil the Lipschitz condition</u> (II.1.2). <u>Let</u> T, w <u>be as in</u> 1., <u>but additionally let</u> $w'(t) \in D(A^{1-\rho})$, $0 \leq t \leq T$, <u>and</u> $A^{1-\rho}w' \in C^0([0,T],B)$. <u>Then the same conclusion as in</u> 1. <u>holds with the exception that the estimate for</u> M(w)' <u>now is</u>

$$\sup_{0\leq s\leq T} \|M(w)'(s)\| \leq k(2 \sup_{0\leq s\leq T} \|Aw(s)\|) \sup_{0\leq s\leq T} \|A^{1-\rho}w'(s)\|.$$

<u>Proof:</u> 1. Let us continue w to an element $w \in C^1([-1,T+1],B)$ with $w(t) \in D(A)$, $-1 \leq t \leq T+1$, $Aw \in C^0([-1,T+1],B)$, $|||w|||_{[-1,T+1]} := \sup_{-1\leq s\leq T+1} \|w'(s)\| + \sup_{-1\leq s\leq T+1} \|Aw(s)\| \leq |||w|||_{[0,T]} := \sup_{0\leq s\leq T} \|w'(s)\| + \sup_{0\leq s\leq T} \|Aw(s)\|$. For the difference quotient

$$\left\| \frac{M(w(s+h))-M(w(s))}{h} \right\| \leq k(2|||w|||_{[0,T]}) \sup_{-1\leq s\leq T+1} \|w'(s)\|,$$

$$h \neq 0, \quad |h| < 1,$$

it follows that for a sequence $h_\nu \to 0$

$$\frac{M(w(.+h_\nu))-M(w(.))}{h_\nu} \to g$$

weak star in $L^\infty((0,T),B)$. Then

$$\lim_{\nu\to\infty} \int_0^T <\frac{M(w(s+h_\nu))-M(w(s))}{h_\nu}, \psi(s)> dt$$

$$= \int_0^T <g(s),\psi(s)> ds$$

$$= \lim_{\nu\to\infty} \int_0^T <M(v(s+h_\nu))-M(v(s)), \frac{1}{h_\nu}[\psi(s)-\psi(s+h_\nu)+\psi(s+h_\nu)]> ds$$

$$= \lim_{\nu\to\infty} \int_0^T \frac{1}{h_\nu}(<M(v(s+h_\nu)),\psi(s+h_\nu)>-<M(v(s)),\psi(s)>) ds +$$

$$+ \lim_{\nu\to\infty} \int_0^T <M(v(s+h_\nu)), \frac{\psi(s)-\psi(s+h_\nu)}{h_\nu}> ds$$

$$= \lim_{\nu\to\infty} \int_0^T \frac{1}{h_\nu}(<M(v(s+h_\nu)),\psi(s+h_\nu)>-<M(v(s)),\psi(s)>) ds -$$

$$- \int_0^T <M(v(s)),\psi'(s)> ds.$$

Using Lebesgue's convergence theorem we see that the first limes is 0.

2. The second part is proved analogously. □

We want to make an important remark: The formula of partial integration holds for $\int_0^t e^{-(t-s)A} M(w(s))\, ds$ if A generates a continuous semigroup. Let us approximate M(w) and M(w)' simultaneously by

$$v_n(t) = \sum_{\nu=1}^{N_n} \vartheta_\nu^{(n)}(t) v_\nu^{(n)}, \quad v_n(0) = M(w(0)),$$

$$v_n'(t) = \sum_{\nu=1}^{N_n} \vartheta_\nu^{(n)\prime}(t) v_\nu^{(n)},$$

$n \to \infty$, in the topology of $L^2((0,T),B)$. Thus $v_n \to M(w)$ in $C^0([0,T],B)$. Here $v_\nu^{(n)} \in B$, $\vartheta_\nu^{(n)} \in C^1([0,T])$. This is possible as it was pointed out in chapter 0. Then it is not difficult to show that

$$\int_0^t e^{-(t-s)A} v_n(s)\, ds = [e^{-(t-s)A} A^{-1} v_n(s)]_0^t - \int_0^t e^{-(t-s)A} A^{-1} v_n'(s)\, ds$$

Letting n tend to ∞ we see that also

$$\int_0^t e^{-(t-s)A} M(w(s))\, ds = [e^{-(t-s)A} A^{-1} M(w(s))]_0^t -$$

$$- \int_0^t e^{-(t-s)A} A^{-1} M(w(s))'\, ds.$$

A similar formula also holds if $e^{-(t-s)A}$ is replaced by $U(t,s)$.

Now we can prove the main result of this paragraph, namely

Theorem II.1.1: Let A be a closed operator in B with domain of definition $D(A)$. Let $\{\lambda \mid \mathrm{Re}\,\lambda \geq 0\}$ be contained in the resolvent set of $-A$. Let $D(A)$ be dense in B. Let $\varphi \in D(A)$.

1. Let $\|(\lambda+A)^{-1}\| \leq \frac{M'}{|\lambda|+1}$ for all λ, $\mathrm{Re}\,\lambda \geq 0$, (then $-A$ generates an analytic semigroup e^{-tA}). Let the mapping $M: D(A) \to B$ fulfil the Lipschitz condition (II.1.2).

Or

2. let $-A$ generate a continuous semigroup e^{-tA} and let the mapping $M: D(A) \to B$ fulfil the Lipschitz condition (II.1.3).

In both cases there exists a quantity $T(\varphi)$, $0 < T(\varphi) \leq +\infty$ _with the following properties: There exists a unique_

$$w \in \bigcap_{0<T<T(\varphi)} C^1([0,T],B)$$

with $w(t) \in D(A)$,

$$Aw \in \bigcap_{0<T<T(\varphi)} C^0([0,T],B)$$

$$w' + Aw + M(w) = 0,$$
$$w(0) = \varphi,$$

$$\lim_{t \uparrow T(\varphi)} \|Aw(t)\| = +\infty, \text{ if } T(\varphi) < +\infty.$$

In the second case we have: $w'(t) \in D(A^{1-\rho})$, $0 < t < T(\varphi)$,

$$t^{1-\rho}A^{1-\rho}w'(t) \in \bigcap_{0<\varepsilon<T<T(\varphi)} C^0([\varepsilon,T],B) \cap \bigcap_{0<T<T(\varphi)} L^\infty((0,T),B).$$

Proof: We start with the part common to both cases. Let $T > 0$, let w be as in lemma II.1.1. We set

$$\mathcal{T}w(t) = e^{-tA}\varphi - \int_0^t e^{-(t-s)A}M(w(s))\, ds.$$

By Lemma II.1.1 we have

$$\mathcal{T}w(t) = e^{-tA}\varphi + \int_0^t e^{-(t-s)A}A^{-1}M(w)'(s)\, ds - [e^{-(t-s)A}A^{-1}M(w(s))]_0^t,$$

$$= e^{-tA}\varphi + \int_0^t e^{-(t-s)A}A^{-1}M(w)'(s)\, ds$$
$$- A^{-1}M(w(t)) + e^{-tA}A^{-1}M(\varphi)$$

$$= e^{-tA}\varphi + e^{-tA}A^{-1}M(\varphi) - A^{-1}M(\varphi) - A^{-1}\int_0^t M(w)'(s)\, ds \tag{II.1.4}$$
$$+ \int_0^t e^{-(t-s)A}A^{-1}M(w)'(s)\, ds.$$

Thus we see that $w \in C^1([0,T],B)$ and

$$\text{(II.1.5)}\quad (\mathfrak{T}w)'(t) = -e^{-tA}A\varphi - e^{-tA}M(\varphi) + \int_0^t e^{-(t-s)A}M(w)'(s)\,ds.$$

We set in the first case

$$C_1 = \sup_{0\le t\le T} \left(2\|e^{-tA}A\varphi\| + 2\|e^{-tA}M(\varphi)\| + 2\|M(\varphi)\|\right)$$

and treat now the first case of our theorem. Let

$$\mathfrak{M}_1 = \{w \mid w \in C^1([0,T],B),\ w(t) \in D(A),\ w(0) = \varphi,$$
$$0 \le t \le T,\ Aw \in C^0([0,T],B),$$
$$|||w|||_{[0,T]} \le C_1 + 1\}.$$

For $w \in \mathfrak{M}_1$ we get with (II.1.4) and (II.1.5) the estimate

$$\|A\mathfrak{T}w(t)\| + \|(\mathfrak{T}w)'(t)\| \le C_1 + Tk(2C_1+2)(C_1+1) +$$
$$+ 2T \sup_{0\le t\le T} \|e^{-tA}\|\cdot k(2C_1+2)(C_1+1).$$

If T is sufficiently small, its size depending on C_1, we thus have shown that

$$\mathfrak{T}(\mathfrak{M}_1) \subset \mathfrak{M}_1.$$

In the second case we set

$$C_1' = \operatorname*{ess\,sup}_{0\le t\le T} 2\left(2\|e^{-tA}A\varphi\| + 2\|e^{-tA}M(\varphi)\| + 2\|M(\varphi)\| +\right.$$
$$\left. + t^{1-\rho}\|A^{1-\rho}e^{-tA}A\varphi\| + t^{1-\rho}\|A^{1-\rho}e^{-tA}M(\varphi)\|\right)$$

and we introduce the set

$$\mathfrak{M}_2 = \{w \mid w \in C^1([0,T],B),\ w'(t) \in D(A^{1-\rho}),$$
$$0 < t \leq T,\ w(t) \in D(A),\ 0 \leq t \leq T,\ w(0) = \varphi,$$
$$A^{1-\rho}w' \in \bigcap_{0<\varepsilon<T} C^0([\varepsilon,T],B),\ t^{1-\rho}A^{1-\rho}w'(.) \in L^\infty((0,T),B)$$
$$Aw \in C^0([0,T],B),$$
$$\operatorname*{ess\,sup}_{0\leq t\leq T} \|t^{1-\rho}A^{1-\rho}w'(t)\| + |||w|||_{[0,T]} \leq C_1'+1\}.$$

Then we have on the basis of (II.1.4), (II.1.5)

$$\operatorname*{ess\,sup}_{0\leq t\leq T} \|t^{1-\rho}A^{1-\rho}w'(t)\| + |||w|||_{[0,T]} \leq$$

$$\leq C_1' + \int_0^T k(2C_1'+2)\cdot\frac{1}{t^{1-\rho}}(C_1'+1)\ dt + 2\int_0^T \frac{1}{t^{1-\rho}}\ dt\ \cdot$$

$$\cdot \sup_{0\leq t\leq T} \|e^{-tA}\| k(2C_1'+2)\cdot(2C_1'+2) + \sup_{0\leq t\leq T} t^{1-\rho}\int_0^t \frac{1}{(t-s)^{1-\rho}}\frac{1}{s^{1-\rho}}\ d$$

$$\sup_{0\leq t\leq T} t^{1-\rho}\|A^{1-\rho}e^{-tA}\| k(2C_1'+2)\cdot(2C_1'+2).$$

Observe that (II.1.4), (II.1.5) are still valid since $M(w)' \in$ $\in \bigcap_{1\leq p<\frac{1}{1-\rho}} L^p((0,T),B)$. Because $t^{1-\rho}\int_0^t \frac{1}{(t-s)^{1-\rho}}\frac{1}{s^{1-\rho}}\ ds \leq t^\rho$

we see that

$$\mathfrak{T}(\mathfrak{M}_2) \subset \mathfrak{M}_2$$

if T is sufficiently small, its size depending on C_1'. Now we set in both cases

$$w_1 = e^{-tA}\varphi,$$
$$w_{n+1} = \mathfrak{T}w_n,\ n \geq 1,$$

and we assume that T is so small that $\mathcal{T}(\mathfrak{M}_1) \subset \mathfrak{M}_1$ and $\mathcal{T}(\mathfrak{M}_2) \subset \mathfrak{M}_2$. Then we get in the first case

$$\|w_{n+2}(t)-w_{n+1}(t)\| = \|\mathcal{T}w_{n+1}(t)-\mathcal{T}w_n(t)\| \leq$$

$$\leq \int_0^t C_1 k(2C_1+2)\|w_{n+1}(s)-w_n(s)\| \; ds$$

and in the second case

$$\|A^{1-\rho}(w_{n+2}(t)-w_{n+1}(t))\| = \|A^{1-\rho}(\mathcal{T}w_{n+1}(t)-\mathcal{T}w_n(t))\| \leq$$

$$\leq \int_0^t \frac{1}{|t-s|^{1-\rho}} C_1' k(2C_1'+2)\|A^{1-\rho}(w_{n+1}(s)-w_n(s))\| \; ds.$$

From this it follows that $w_n \to w$ in $C^0([0,T],B)$ and $A^{1-\rho}w_n \to A^{1-\rho}w$ in $C^0([0,T],B)$ respectively. Our Lipschitz condition on M then shows that

$$M(w_n) \to M(w) \text{ in } C^0([0,T],B)$$

in both cases. Since

$$w'_{n+1} + Aw_{n+1} + M(w_n) = 0$$

$$w_{n+1}(0) = \varphi$$

we can choose a subsequence w_{j_n} with

$$w'_{j_n} \to w' \text{ weak star in } L^\infty((0,T),B),$$

$$Aw'_{j_n} \to Aw \text{ weak star in } L^\infty((0,T),B)$$

in the first case and

$$t^{1-\rho}A^{1-\rho}w'_{j_n}(t) \to t^{1-\rho}A^{1-\rho}w'(t) \quad \text{weak star in } L^\infty((0,T),B)$$

$$Aw_{j_n} \to Aw \quad \text{weak star in } L^\infty((0,T),B)$$

in the second case. In both cases

$$w' + Aw + M(w) = 0,$$
$$w(0) = \varphi.$$

Moreover, the limit process above shows that (II.1.4) holds for w instead of $\mathcal{T}w$ on the left hand side. Thus $Aw \in C^0([0,T],B)$, $w \in C^1([0,T],B)$. From (II.1.5) it follows that $t^{1-\rho}A^{1-\rho}w'(t) \in \in \bigcap_{0<\varepsilon\leq T} C^0([\varepsilon,T],B) \cap L^\infty((0,T),B)$. For two solutions $w,\tilde{w}$ on $[0,T]$ of $u' + Aw + M(w) = 0$, $w(0) = \tilde{w}(0) = \varphi$, as before we have

$$\|w(t)-\tilde{w}(t)\| \leq \int_0^t \sup_{0\leq s\leq t} \|e^{-sA}\| k(C)\|w(s)-\tilde{w}(s)\| \, ds$$

with $C \geq \sup_{0\leq s\leq t} (\|Aw(s)\| + \|A\tilde{w}(s)\|)$, $0 \leq t \leq T$. Therefore $w(t) = \tilde{w}(t)$ on $[0,T]$. Set $T_1 = T$. We can continue our construction on a second small intervall $[T_1,T_2]$ and so on. This construction ends below a number $T(\varphi) > 0$ if $\lim_{t\uparrow T(\varphi)} \|Aw(t)\| \to \infty$, otherwise w can be continued to the whole real axis. □

Finally we briefly discuss the time dependent case $u' + A(t)u + M(u) = 0$, $u(0) = \varphi$. We restrict ourselves to the parabolic case, i.e. each $-A(t)$ generates an analytic semigroup $e^{-\tau A(t)}$, $\tau \geq 0$. Let $T > 0$. We assume that the $A(t)$, $0 \leq t \leq T$, fulfill (U1)-(U3) in I, § 2.

Then we can construct the evolution operator $U(t,s)$. The nonlinearity M is defined as a mapping from $D(A(0))$ into B fulfilling the following relations:

(II.1.6) $\|M(u)\| \leq k(C)$, $\|M(u)-M(v)\| \leq k(C)\|A^{1-\rho}(t)(u-v)\|$, for some $\rho \in (0,1)$, $u,v \in D(A(0))$, $C \geq \|A(0)u\|+\|A(0)v\|$. Observe that instead of $C \geq \|A(0)u\|+\|A(0)v\|$ one also can prescribe $C \geq \|A(t_o)u\|+\|A(t_o)v\|$ for some $t_o \in [0,T]$; moreover $A^{1-\rho}(t)$ may be replaced by $A^{1-\rho+\varepsilon}(t_o)$ for some $t_o \in [0,T]$ and some $\varepsilon \in (0,1)$ with $1-\rho+\varepsilon < 1$.

Now using (I.2.3) we can prove again an existence theorem for strong solutions of nonlinear differential equations of parabolic type in the Banach space B, namely

Theorem II.1.2: *Let* $\varphi \in D(A)$. *Then there exists a positive* $T(\varphi) \leq T$ *having the following properties*: *There exists a unique element*

$$u \in \bigcap_{0<\tilde{T}<T(\varphi)} C^1([0,\tilde{T}],B) \text{ with}$$

$$u(t) \in D(A(0)),\ 0 \leq t < T(\varphi),\ u'(t) \in D(A^{1-\rho}(t)),\ 0 < t < T(\varphi),$$

$$A(.)u(.) \in \bigcap_{0<\tilde{T}<T(\varphi)} C^0([0,\tilde{T}],B),$$

$$\cdot^{1-\rho}A^{1-\rho}(.)u'(.) \in \bigcap_{0<\tilde{T}<T(\varphi)} C^0((0,\tilde{T}],B) \cap L^\infty((0,\tilde{T}),B),$$

$$u' + A(t)u + M(u) = 0,$$
$$u(0) = \varphi.$$

If $T(\varphi) < T$ *then*

$$\lim_{t \uparrow T(\varphi)} \|A(t)u(t)\| = \infty.$$

If $T(\varphi) = T$ *and* $\sup_{0 \leq t<T} \|A(t)u(t)\| < \infty$ *then* u *may be continued on* $[0,T]$ *as to be an element of* $C^1([0,T],B)$ *with* $u(t) \in D(A(0))$, $0 \leq t \leq T$, $A(.)u(.) \in C^0([0,T],B)$, $u'(t) \in D(A^{1-\rho}(t))$, $0 < t \leq T$, $\cdot^{1-\rho}A^{1-\rho}(.)u'(.) \in C^0((0,T],B) \cap L^\infty((0,T),B)$, $u'+A(t)u+M(u) = 0$, $0 \leq t \leq T$, $U(0) = \varphi$.

It is of course allowed that the nonlinearity depends on t too. In this case the Lipschitz condition on M reads as follows: For each t, $t \in [0,T]$, M is a mapping from $D(A(0))$ into B satisfying the following Lipschitz condition:

$$\|M(t,u)-M(s,v)\| \leq k(C)(|t-s| + \|A^{1-\rho}(t)(u-v)\|), \quad 0 \leq t,s \leq T,$$

where $C \geq \|A(0)u\| + \|A(0)v\| + t + s$. The following result then may be considered as a

<u>Corollary to Theorem II.1.2:</u> <u>Let</u> $\varphi \in D(A)$. <u>Then there exists a positive number</u> $T(\varphi) \leq T$ <u>having the following properties</u>: <u>There exists a unique element</u>

$$u \in \bigcap_{0<\tilde{T}<T(\varphi)} C^1([0,\tilde{T}],B) \quad \underline{\text{with}}$$

$$u(t) \in D(A(0)), \ 0 \leq t < T(\varphi), \quad u'(t) \in D(A^{1-\rho}(t)), \ 0 < t < T(\varphi),$$

$$A(.)u(.) \in \bigcap_{0<\tilde{T}<T(\varphi)} C^0([0,\tilde{T}],B),$$

$$.^{1-\rho}A^{1-\rho}(.)u'(.) \in \bigcap_{0<\tilde{T}<T(\varphi)} C^0((0,\tilde{T}],B) \cap L^\infty((0,\tilde{T}),B)$$

$$u' + A(t)u + M(t,u) = 0,$$
$$u(0) = \varphi.$$

<u>If</u> $T(\varphi) < T$ <u>then</u>

$$\lim_{t \uparrow T(\varphi)} \|A(t)u(t)\| = \infty.$$

<u>If</u> $T(\varphi) = T$ <u>and</u> $\sup_{0 \leq t < T} \|A(t)u(t)\| < \infty$ <u>then</u> u <u>may be continued on</u> $[0,T]$ <u>as to be an element with the same properties as above on</u> $[0,T]$ <u>instead of</u> $[0,\tilde{T}]$, $0 < \tilde{T} < T(\varphi)$, <u>and on</u> $(0,T]$, $(0,T)$ <u>instead of</u> $(0,\tilde{T}]$, $(0,\tilde{T})$ <u>resp</u>.

§ 2. A Nonlinear Interpolation Theorem

In this paragraph we study the concept of an analytic mapping $f:B_1 \to B_2$ between two Banach spaces B_1, B_2. Our notion of analyticity is that given in [HP] and will be explained in what follows. One reason for doing so is that e.g. analytic nonlinearities M in the equations treated in § 1 allow us to prove higher regularity for the solutions constructed in § 1 on their interval of existence $[0,T(\varphi))$. Another is that the nonlinearity in the Navier-Stokes equations is an analytic mapping for suitable B_i.
Thus we use the notion of an analytic mapping as a tool for our proofs but we do not claim that the corresponding theorems could not be proved otherwise. Our access is a generalization of the well known complex interpolation method in the linear case.

Definition II.2.1: Let B_1, B_2 two complex Banach spaces with norms $\|\cdot\|_{B_k}$, $k = 1,2$. Then $A(B_1,B_2)$ is the class of all mappings $f:B_1 \to B_2$ with the following properties:

a) On every ball $\{x \mid \|x\|_{B_1} \leq R\}$ with $R < +\infty$ the images $f(x)$ are bounded; i.e. there is a monotone nondecreasing continuous function $\omega:[0,\infty) \to [0,+\infty)$ such that

$$\text{(II.2.1)} \quad \|f(x)\|_{B_2} \leq \omega(\|x\|_{B_1}), \quad x \in B_1.$$

b) For $x,y \in B_1$ the mapping

$$\zeta \mapsto f(x+\zeta\eta)$$

is holomorphic from $\mathbb{C}$ into B_2. Holomorphic of course means complex differentiable.

Cauchy's formula in a complex neighbourhood $U(\zeta_o)$ of ζ_o:

$$f(x+\zeta\eta) = -\frac{1}{2\pi i}\int_{\partial K_1(\zeta_o)} \frac{f(x+z\eta)}{\zeta-z}\,dz,\ \zeta\in U(\zeta_o),$$

$K_1(\zeta_o) = \{z \mid |z-\zeta_o| \leq 1\}$, then shows that

$$\left\|\frac{df}{d\zeta}(x+\zeta\eta)\right\|_{B_2} \leq \omega(\|x\|_{B_1} + \|\eta\|_{B_1} + |\zeta|\,\|\eta\|_{B_1})$$

and consequently

$$\frac{\|f(x')-f(x'')\|_{B_2}}{\|x''-x'\|_{B_1}} = \frac{\|f(x')-f(x'+\|x''-x'\|_{B_1}\frac{x''-x'}{\|x''-x'\|_{B_1}})\|_{B_2}}{\|x''-x'\|_{B_1}}$$

$$\leq \omega(\|x'\|_{B_1} + \|x''-x'\|_{B_1} + 1).$$

Thus f is Lipschitz continuous.

Now we assume that the linear operators $A(t)$, $0 \leq t \leq T$, in the Banach space B fulfill the assumptions (U1)-(U3) of I, § 2. Then we can define the fractional powers $A^\alpha(t)$, $\operatorname{Re}\alpha > 0$. $D(A^\alpha(t))$ is a Banach space with norm $\|x\|_{D(A^\alpha(t))} = \|A^\alpha(t)x\|$. Thus the class $A(D(A^{\alpha_1}(t)), D(A^{\alpha_2}(t)))$ is well defined. We already studied in I, § 3 the holomorphy of $A^\zeta(t)e^{-\tau A(t)}$ with respect to ζ, where $\tau > 0$, $0 \leq t \leq T$. If f is holomorphic from $\{\zeta \mid \operatorname{Re}\zeta > 0\}$ into B and if $g \in A(B,B)$ then $g \circ f$ is holomorphic from $\{\zeta \mid \operatorname{Re}\zeta > 0\}$ into B. The proof is left to the reader (see e.g. [HP, p. 112]).

We want to prove the following

Theorem II.2.1: Besides our assumptions on the $A(t)$ above the $A(t)$ are supposed to have the following closedness property: Let $\delta \geq 0$. Let $x_n \to x$ in B, $\|A^\delta(t)x_n\| \leq C$ for any $t \in [0,T]$. Then $x \in D(A^\delta(t))$ and $\|A^\delta(t)x\| \leq C$.

Let $\alpha_1,\alpha_2,\beta_1,\beta_2 \geq 0$, $\alpha_2 > \alpha_1 > 0$, $\beta_2 > 0$. _Let_

$$f \in A(D(A^{\alpha_1}(t)),D(A^{\beta_1}(t))),$$

$$f \in A(D(A^{\alpha_2}(t)),D(A^{\beta_2}(t))),\quad 0 \leq t \leq T.$$

Then $f(D(A^{\gamma+\eta}(t)) \subseteq D(A^{\delta-\eta}(t))$, _with_ $\eta > 0$,

$$\gamma = \sigma\alpha_1 + (1-\sigma)\alpha_2,\quad \gamma+\eta \leq \alpha_2,$$

$$\delta = \sigma\beta_1 + (1-\sigma)\beta_2,\quad \sigma \in [0,1]$$

and

$$\|A^{\delta-\eta}(t)f(u)\| \leq \omega_\eta(\|A^{\gamma+\eta}(t)u\|),\quad t \in [0,T],\quad u \in D(A^{\gamma+\eta}(t)u).$$

ω_η _is a monotone nondecreasing continuous function from_ $[0,\infty)$ _into itself depending on_ $\eta > 0$.

Proof: According to our assumptions we have

$$\|A^{\beta_k}(t)f(x)\| \leq \omega(\|A^{\alpha_k}(t)x\|),\quad x \in D(A^{\alpha_k}(t)),\quad k = 1,2,\quad t \in [0,T].$$

Let $\varepsilon,\rho > 0$. For $x \in B$ we set

$$\varphi(\sigma) = A^{\sigma\beta_1+(1-\sigma)\beta_2}(t)e^{-A(t)\rho}f(A^{-(\sigma\alpha_1+(1-\sigma)\alpha_2)}(t)e^{-A(t)\varepsilon}x),$$

$0 \leq \mathrm{Re}\ \sigma \leq 1$. As it was proved in I, § 3 the mapping

$$\sigma \mapsto A^{-(\sigma\alpha_1+(1-\sigma)\alpha_2)}(t)e^{-A(t)\varepsilon}x$$

is holomorphic from $\{\sigma | 0 < \mathrm{Re}\ \sigma < 1\}$ into $D(A^{\alpha_k}(t))$ and continuous from $\{\sigma | 0 \leq \mathrm{Re}\ \sigma \leq 1\}$ into $D(A^{\alpha_k}(t))$, $k = 1,2$. Therefore

$$\sigma \mapsto f(A^{-(\sigma\alpha_1+(1-\sigma)\alpha_2)}(t)e^{-A(t)\varepsilon}x)$$

is holomorphic from $\{\sigma | 0 < \mathrm{Re}\ \sigma < 1\}$ into B (observe that

$f \in A(D(A^{\alpha_k}(t)), D(A^{\beta_k}(t)))!)$ and continuous from $\{\alpha | 0 \le \mathrm{Re}\ \sigma \le 1\}$ into B. Thus φ is holomorphic from $\{\sigma | 0 < \mathrm{Re}\ \sigma < 1\}$ into B and continuous from $\{\sigma | 0 \le \mathrm{Re}\ \sigma \le 1\}$ into B if $\beta_1 > 0$. If $\beta_1 = 0$ then φ is holomorphic from $\{\sigma | 0 < \mathrm{Re}\ \sigma < 1\}$ into B and continuous from $\{\sigma | 0 \le \mathrm{Re}\ \sigma < 1\}$ into B. If $1 > \mathrm{Re}\ \zeta > 0$ then it easily follows from our representation for $A^{-\alpha}(t)$ in I, § 3, $\mathrm{Re}\ \alpha > 0$, that

$$A^{\zeta}(t)A^{-1}(t)x = \frac{\sin \zeta\pi}{\pi} \int_0^{\infty} \lambda^{\zeta-1}(\lambda+A(t))^{-1}x\, d\lambda,\ x \in B,$$

(see [Y2, p. 260] for this formula for fractional powers). Thus

$$A^{\zeta}(t)A^{-1}(t)x = \frac{\sin \zeta\pi}{\pi} \left\{\int_0^1 \lambda^{\zeta-1}((\lambda+A(t))^{-1}-A^{-1}(t))x\, d\lambda + \right.$$
$$\left. + \int_0^1 \lambda^{\zeta-1}A^{-1}(t)x\, d\lambda\right\} +$$
$$+ \frac{\sin \zeta\pi}{\pi} \int_1^{\infty} \lambda^{\zeta-1}(\lambda+A(t))^{-1}x\, d\lambda,\ x \in B.$$

In virtue of the first resolvent equation the first integral and the last one can be continued continuously on $\mathrm{Re}\ \zeta = 0$. Since the second one fulfils

$$\frac{\sin \zeta\pi}{\pi} \int_0^1 \lambda^{\zeta-1}A^{-1}(t)x\, d\lambda = \frac{\sin \zeta\pi}{\zeta\pi} A^{-1}(t)x$$

we see that $A\ (t)A^{-1}(t)x$ can be continued continuously on $\mathrm{Re}\ \zeta = 0$. Writing $A^{(1-\sigma)\beta_2}(t)A^{-1}(t)A(t)e^{-A(t)\rho}$ instead of $A^{(1-\sigma)\beta_2}(t)e^{-A(t)\rho}$ we see that $\varphi(\sigma)$ can be continued continuously on $\mathrm{Re}\ \sigma = 1$ if $\beta_1 = 0$. As we have proved in I, § 3 we have $(M \in \mathbb{N})$

$$\|A^{-\zeta}(t)\| \le \frac{c}{|\Gamma(\zeta)\ \mathrm{Re}\ \zeta|},\ 0 < \mathrm{Re}\ \zeta,$$

$$\|A^{\zeta}(t)e^{-A(t)\tau}\| \le \frac{\tilde{c}e^{-c\tau}}{(1-\mathrm{Re}\zeta)\tau},\ 0 < \mathrm{Re}\ \zeta < 1,$$

$$\|A^{\zeta}(t)e^{-A(t)\tau}\| \le \prod_1^M \|A^{\frac{\zeta}{M}}(t)e^{-\frac{\tau}{M}A(t)}\| \le \frac{\tilde{c}e^{-c\tau}}{(1-\frac{\mathrm{Re}\ \zeta}{M})^M (\frac{\tau}{M})^M},\ 0 < \mathrm{Re}\ \zeta < M.$$

Thus the expressions $\|A^{\zeta}(t)e^{-\tau A(t)}\|$ remain uniformly bounded in $0 < \mathrm{Re}\ \zeta \le \frac{M}{2}$, and φ is uniformly bounded in $0 \le \mathrm{Re}\ \zeta \le 1$. We are now in a position to apply Hadamard's Three Lines Theorem on φ. Then we have

$$\|A^{-\eta}\varphi(\sigma)\|. \le \Big(\sup_{-\infty<t<\infty} \|A^{\beta_2}(t)e^{-A(t)\rho} f(A^{-\alpha_2+i\tilde{t}(\alpha_2-\alpha_1)-\eta} e^{-A(t)\varepsilon} A^{\eta}(t)x)\| \Big)^{1-\mathrm{Re}\ \sigma} .$$

$$\cdot \frac{c}{\eta\Gamma(\eta)} \Big(\sup_{-\infty<t<\infty} \|A^{\beta_1}(t)e^{-A(t)\rho} f(A^{-\alpha_1+i\tilde{t}(\alpha_2-\alpha_1)-\eta} e^{-A(t)\varepsilon} A^{\eta}(t)x)\| \Big)^{\mathrm{Re}\ \sigma},$$

$$\le \omega_{\eta}(\|A^{\eta}(t)x\|), \quad x \in D(A^{\eta}(t)).$$

Let $u \in D(A^{\gamma+\eta}(t))$, $x = A^{\gamma}(t)u$. Then it follows that

$$\|A^{\delta-\eta}(t)e^{-A(t)\rho} f(e^{-A(t)\varepsilon}u)\| \le \omega_{\eta}(\|A^{\gamma+\eta}(t)u\|).$$

Letting ρ tend to 0 we get by our closedness assumption:
$f(e^{-A(t)\varepsilon}u) \in D(A^{\delta-\eta}(t))$

$$\|A^{\delta-\eta}(t)f(e^{-A(t)\varepsilon}u)\| \le \omega_{\eta}(\|A^{\gamma+\eta}(t)u\|).$$

Since $D(A^{\gamma}(t)) \subset D(A^{\alpha_1}(t))$ and $e^{-A(t)\varepsilon}u \to u$ in $D(A^{\alpha_1}(t))$ for $\varepsilon \to 0$ we have

$$f(e^{-A(t)\varepsilon}u) \to f(u)$$

in B for $\varepsilon \to 0$. Applying again our closedness assumption we get:
$f(u) \in D(A^{\delta-\eta}(t))$,

$$\|A^{\delta-\eta}(t)f(u)\| \le \omega_{\eta}(\|A^{\gamma+\eta}(t)u\|).$$

We also want to give two additional versions of Theorem II.2.1 where we specialise the function ω and consider the influence of such a specialization on the intermediate estimates.

Theorem II.2.2: Let all the assumptions of Theorem II.2.1 be fulfilled. Moreover let $\alpha_2 \geq \beta_2$. Let

$$\|A^{\beta_1}(t)f(x)\| \leq \omega_1(\|A^{\alpha_1}(t)x\|), \ x \in D(A^{\alpha_1}(t)), \ t \in [0,T],$$

$$\|A^{\beta_2}(t)f(x)\| \leq \omega_2(\|A^{\beta_2-\varepsilon}(t)x\|)\|A^{\alpha_2}(t)x\|, \ x \in D(A^{\alpha_2}(t)), \ t \in [0,T],$$

for some $\varepsilon > 0$. ω_1, ω_2 are as ω above. Then

$$f(D(A^{\gamma+\eta}(t))) \subset D(A^{\delta-\eta}(t)), \ \eta > 0,$$

$$\gamma = \sigma\alpha_1 + (1-\sigma)\alpha_2, \ \gamma+\eta \leq \alpha_2,$$

$$\delta = \sigma\beta_1 + (1-\sigma)\beta_2, \ \sigma \in [0,1]$$

and

$$\|A^{\delta-\eta}(t)f(x)\| \leq \omega_{2\eta}(\|A^{\gamma+\eta-(\alpha_2-\beta_2)-\varepsilon}(t)x\|)^{1-\sigma} \|A^{\gamma+\eta}(t)x\|^{1-\sigma}\omega_{1\eta}(\|A^{\gamma+\eta}(t)x\|)^{\sigma},$$

$$x \in D(A^{\gamma+\eta}(t)), \ t \in [0,T],$$

with $\omega_{2\eta}, \omega_{1\eta}$ as ω_η above.

Proof: The proof is the same as that one of Theorem II.2.1 until the application of Hadamard's Three Lines Theorem. The application of the Three Lines Theorem here furnishes the estimate

$$\|A^{-\eta}(t)\varphi(\sigma)\| \leq \|A^{\eta}(t)x\|^{1-\mathrm{Re}\,\sigma}\omega_{2\eta}(\|A^{\beta_2+\eta-\alpha_2-\varepsilon}(t)x\|)^{1-\mathrm{Re}\,\sigma} \cdot \omega_{1\eta}(\|A^{\eta}(t)x\|)^{\mathrm{Re}\,\sigma}.$$

Let $u \in D(A^{\gamma+\eta}(t))$, $x = A^{\gamma}(t)u$. Then

$$\|A^{\delta-\eta}(t)e^{-A(t)\rho}f(e^{-A(t)\varepsilon}u)\| \leq$$

$$\leq \|A^{\gamma+\eta}(t)u\|^{1-\sigma}\omega_{2\eta}(\|A^{\gamma+\eta-(\alpha_2-\beta_2)-\varepsilon}(t)u\|)^{1-\sigma}\cdot\omega_{1\eta}(\|A^{\gamma+\eta}(t)u\|)^{\sigma}.$$

From this our assertion follows as in the proof of Theorem II.2.1.

Next we need a refinement of Theorem II.2.2. This we get by specialising ω_1. We omit the proof since it is very similar to that of Theorem II.2.1 or II.2.2.

<u>Theorem II.2.3:</u> <u>Let the assumptions of Theorem</u> II.2.2 <u>be fulfilled for all</u> $t \in [0,T]$. <u>Let</u>

$$\|A^{\beta_1}(t)f(x)\| \leq \tilde{\omega}_1(\|A^{\alpha_1-\varepsilon'}(t)x\|)\|A^{\alpha_1}(t)x\|, \quad x \in D(A^{\alpha_1}(t)),$$

$$\|A^{\beta_2}(t)f(x)\| \leq \tilde{\omega}_2(\|A^{\beta_2-\varepsilon'}(t)x\|)\|A^{\alpha_2}(t)x\|, \quad x \in D(A^{\alpha_2}(t)),$$

<u>with monotone non decreasing functions</u> $\tilde{\omega}_1, \tilde{\omega}_2 : [0,+\infty) \to [0,+\infty)$ <u>and some</u> $\varepsilon' > 0$. <u>Then again</u> $f(A^{\gamma+\eta}(t)) \subset D(A^{\delta-\eta}(t))$ <u>with</u> $\eta > 0$,

$$\gamma = \sigma\alpha_1 + (1-\sigma)\alpha_2, \quad \gamma+\eta \leq \alpha_2,$$

$$\delta = \sigma\beta_1 + (1-\sigma)\beta_2, \quad \sigma \in [0,1]$$

<u>and</u>

$$\|A^{\delta-\eta}(t)f(x)\| \leq \tilde{\omega}_{2\eta}(\|A^{\gamma+\eta-(\alpha_2-\beta_2)-\varepsilon'}(t)x\|)^{1-\sigma}\cdot$$

$$\cdot\tilde{\omega}_{1\eta}(\|A^{\gamma+\eta-\varepsilon'}(t)x\|)^{\sigma}\|A^{\gamma+\eta}(t)x\|,$$

$$x \in D(A^{\gamma+\eta}(t)), \quad t \in [0,T]$$

<u>with</u> $\tilde{\omega}_{1\eta}$, $\tilde{\omega}_{2\eta}$ <u>as</u> $\omega_{1\eta}$, $\omega_{2\eta}$ <u>in Theorem</u> II.2.2.

Remark: If B is a Hilbert space and if the A(t) are selfadjoint then η is allowed to be 0 in Theorems II.2.1 - II.2.3; see [H2] and "Comments to chapter II, § 2".

§ 3. Solutions of Equations with Nonlinearities Relatively Bounded to $A^{1-\rho}$, their Higher Regularity and the Question of Admissible Initial Data

In this paragraph we study equations of the same form as in II., § 1, but we assume that M is a mapping from $D(A^{1-\rho})$ into B for some $\rho \in (0,1)$. B does not need to be reflexive. We have two different cases, namely: The initial value fulfills high regularity assumptions or the initial value is simply from B or $D(A^{\delta})$ for a small δ. The case of t-dependence of A, i.e. the equation $u' + A(t)u = M(u)$, will be briefly discussed in the comments to II, § 3. A is always supposed to fulfill the closedness property of theorem II.2.1.

<u>Theorem II.3.1:</u> <u>Let</u> $\rho_1, \rho_2 \in (0,1)$, <u>let</u> $\kappa \geq 1$. <u>Let</u>

$$M: D(A^{1-\rho_1}) \to B$$

<u>be from</u> $A(D(A^{1-\rho_1}), B)$ <u>and let</u>

$$M: D(A^{\kappa}) \to D(A^{\kappa-(1-\rho_2)})$$

<u>be from</u> $A(D(A^{\kappa}), D(A^{\kappa-(1-\rho_2)}))$. <u>Let</u> $\varphi \in D(A^{\kappa})$. <u>Then there exists</u> <u>a</u> T_φ, $0 < T_\varphi \leq +\infty$, <u>with the following property</u>: <u>There is exactly</u> <u>one</u> $u: [0,\tilde{T}] \to D(A)$, $\tilde{T} < T_\varphi$, <u>with</u>

$$(\text{II.3.1}) \quad \begin{cases} Au(.) \in \bigcap_{0<\tilde{T}<T_\varphi} C^0([0,\tilde{T}],B), \\ \quad u \in \bigcap_{0<\tilde{T}<T_\varphi} C^1([0,\tilde{T}],B), \\ u' + Au + M(u) = 0, \\ \quad u(0) = \varphi, \\ \lim_{t \uparrow T_\varphi} \|A^{1-\rho}u(t)\| = \infty \ \underline{\text{if}}\ T_\varphi < +\infty. \end{cases}$$

Moreover this solution has the following regularity properties:
$u:[0,\tilde{T}] \to D(A^{\kappa})$, $\tilde{T} < T_{\varphi}$,

$$A^{\kappa}u(.) \in \bigcap_{0<\tilde{T}<T_{\varphi}} C^{0}([0,\tilde{T}],B),$$

$$u'(t) \in D(A^{\kappa-1}),\ 0 \leq t \leq \tilde{T},$$

$$A^{\kappa-1}u'(.) \in \bigcap_{0<\tilde{T}<T_{\varphi}} C^{0}([0,\tilde{T}],B).$$

For the first part of the theorem we only need the Lipschitz continuity of the mapping $M:D(A^{1-\rho_1}) \to B$. This follows from the analyticity (see § 2). If there exists for φ a continuous function $f_{\varphi}:[0,\infty) \to [0,\infty)$ with $\|A^{1-\rho}u(t)\| \leq f_{\varphi}(t)$ on every interval of existence $[0,\tilde{T}]$ of u, where u fulfills (II.3.1), then $T_{\varphi} = +\infty$.

Proof: As the Picard-iteration shows the integral equation

$$\text{(II.3.2)} \quad u(t) = e^{-tA}\varphi - \int_{0}^{t} e^{-(t-s)A} M(u(s))\, ds$$

has a unique solution u on a maximal interval $[0,T_{\varphi})$; this means that there exists a mapping $u:[0,T_{\varphi}) \to D(A^{1-\rho})$ with

$$A^{1-\rho}u \in \bigcap_{0<\tilde{T}<T_{\varphi}} C^{0}([0,\tilde{T}],B)$$

which fulfills (II.3.2). If $T_{\varphi} < +\infty$ then

$$\lim_{t \uparrow T_{\varphi}} \|A^{1-\rho}u(t)\| = \infty.$$

Now we want to show that $A^{1-\rho}u$ is even Hölder continuous.
We have

$$\text{(II.3.3)}\quad \|A^{\gamma}(e^{-(t+\Delta t-s)A}-e^{-(t-s)A})x\|$$

$$\le \|A^{\gamma}\int_0^{\Delta t} Ae^{-(t+\sigma-s)A}x\,d\sigma\|,$$

$$\le \int_0^{\Delta t}\frac{c}{|t+\sigma-s|^{1+\gamma}}\,d\sigma \le \left|\frac{c}{|t+\Delta t-s|^{\gamma}}-\frac{c}{|t-s|^{\gamma}}\right|,$$

$$\le \frac{c}{|t-s|^{\gamma}}\;\frac{\big||t-s|^{\gamma}-|t+\Delta t-s|^{\gamma}\big|}{|t+\Delta t-s|^{\gamma}},$$

$$\le \frac{c|\Delta t|^{\gamma}}{|t-s|^{\gamma}|t+\Delta t-s|^{\gamma}} \le \frac{c|\Delta t|^{\delta}}{|t-s|^{\gamma+\delta}},$$

$\gamma \ge \delta \ge 0$, $\Delta t \ge 0$, $t > s$. Thus on $[0,\tilde{T}]$

$$\|A^{1-\rho}\int_0^{t+\Delta t} e^{-(t+\Delta t-s)A}M(u(s))\,ds - A^{1-\rho}\int_0^{t} e^{-(t-s)A}M(u(s))\,ds\|$$

$$\le (c(\rho)\Delta t^{\rho}+c(\rho)|\Delta t|^{\delta}t^{\rho-\delta})\cdot g(\sup_{0\le s\le t}\|A^{1-\rho}u(s)\|),$$

with some continuous function $g:[0,\infty)\to[0,\infty)$ and for all $\delta\in[0,\rho)\cap[0,1-\rho)$. Moreover we have

$$\|A^{1-\rho}(e^{-(t+\Delta t)A}-e^{-tA})\varphi\|$$

$$\le \|(e^{-(t+\Delta t)A}-e^{-tA})A^{1-\rho-\kappa}\|\,\|A^{\kappa}\varphi\|,$$

$$\le \int_0^{\Delta t}\|A^{2-\rho-\kappa}e^{-(t+\sigma)A}\|\,d\sigma\,\|A^{\kappa}\varphi\|,$$

$$\le c|\Delta t|^{\rho+\kappa-1}\|A^{\kappa}\varphi\|.$$

It may be left to the reader to prove the corresponding inequalities for $\Delta t \le 0$. Thus we arrive at the conclusion that

$$A^{1-\rho}u\in C^{\varepsilon}([0,\tilde{T}],B),\quad \tilde{T}<T_{\varphi},$$

for some $\varepsilon=\varepsilon(\rho,\kappa)>0$. Using the Lipschitz continuity of M we get with the method employed in the proof of theorem I.2.1 the relations (II.3.1). For the second part of our theorem we want to apply theorem II.2.1. We set

$$\alpha_1 = 1-\rho_1, \quad \beta_1 = 0,$$
$$\alpha_2 = \kappa, \qquad \beta_2 = \kappa-(1-\rho_2).$$

Then we have

$$\gamma = \sigma\alpha_1+(1-\sigma)\alpha_2, \quad \gamma+\eta \leq \alpha_2,$$
$$\delta = \sigma\beta_1+(1-\sigma)\beta_2, \quad 0 \leq \sigma \leq 1,$$

i.e.

$$\gamma = \sigma(1-\rho_1) + (1-\sigma)\kappa, \quad \gamma+\eta \leq \kappa,$$
$$\delta = (1-\sigma)(\kappa-(1-\rho_2)), \quad 0 \leq \sigma \leq 1.$$

Then we have for $\varepsilon,\eta \in (0,1)$:

$$\text{(II.3.4)} \quad \delta - \eta + 1 - \varepsilon - (\gamma+\eta) = 1 - \varepsilon - \sigma(1-\rho_1) - (1-\sigma)(1-\rho_2)^{-2\eta},$$
$$\geq 1 - \varepsilon - (1 - \min(\rho_1,\rho_2))^{-2\eta},$$
$$= \min(\rho_1,\rho_2) - \varepsilon - 2\eta > 0$$

if $\min(\rho_1,\rho_2) > \varepsilon+2\eta > 0$. For a moment we assume that we already know: $u(t) \in D(A^{\gamma+\eta})$, $0 \leq t \leq \tilde{T}$, $A^{\gamma+\eta}u(.) \in C^0([0,\tilde{T}],B)$ for some $\tilde{T} \in (0,T(\varphi))$. Then according to theorem II.2.1

$$A^{\delta-\eta+1-\varepsilon}u(t) = e^{-tA}A^{\delta-\eta+1-\varepsilon}\varphi - \int_0^t A^{1-\varepsilon}e^{-(t-s)A}A^{\delta-\eta}M(u(s))\,ds,$$

$$\|A^{1-\varepsilon}e^{-tA}\| \leq \frac{c}{|t-s|^{1-\varepsilon}},$$

$$\|A^{\delta-\eta}M(u(s))\| \leq w_\eta(\|A^{\gamma+\eta}u(s)\|)$$

if $\delta-\eta+1-\varepsilon \leq \kappa$. Thus (II.3.4) shows that the regularity of u is improved stepwise at least by $\min(\rho_1,\rho_2) - \varepsilon - 2\eta$ if one already knows that $u(t) \in D(A^{1-\rho_1+\eta})$, $0 \leq t \leq \tilde{T}$, $A^{1-\rho_1+\eta}u(.) \in C^0([0,\tilde{T}],B)$. This we have just proved; therefore we arrive after finitely many steps at the desired regularity result.

Next we study the case that the initial value φ is only from $D(A^{\kappa-(1-\rho_2)})$ instead of $D(A^{\kappa})$.

<u>Theorem II.3.2: Let</u> ρ_1,ρ_2,κ,M <u>be as in the preceding theorem. Let</u> $\varphi \in D(A^{\kappa-(1-\rho_2)}) \cap D(A^{1-\rho_1})$. <u>Then there exists a</u> T_φ, $0 < T_\varphi \leq +\infty$ <u>with the following property: There is exactly one</u> $u:[0,\tilde{T}] \to$ $\to D(A^{1-\rho_1})$, $\tilde{T} < T_\varphi$, <u>with</u>

$$A^{1-\rho}u(.) \in \bigcap_{0<\tilde{T}<T_\varphi} C^0([0,\tilde{T}],B),$$

$$u(t) = e^{-tA}\varphi - \int_0^t e^{-(t-s)A}M(u(s))\,ds.$$

<u>Moreover</u> $\frac{d^\nu}{dt^\nu}u(t) \in D(A^{\kappa-(1-\rho_2)+1-\varepsilon-\nu})$, $0 < t \leq \tilde{T}$, $0 < \varepsilon \leq 1$, $\nu = 0,1$, <u>and</u>

$$A^{\kappa-(1-\rho_2)+1-\varepsilon-\nu}\frac{d^\nu}{dt^\nu}u(.) \in \bigcap_{0<\tilde{T}<T_\varphi} C^0((0,\tilde{T}],B).$$

<u>Proof:</u> The integral equation can be solved as in the proof of the preceding theorem. Also as in the proof of the preceding theorem we get

$$u(t) \in D(A^{\kappa-(1-\rho_2)}),$$

$$A^{\kappa-(1-\rho_2)}u(.) \in \bigcap_{0<\tilde{T}<T_\varphi} C^0([0,\tilde{T}],B);$$

namely, for

$$\gamma = \sigma(1-\rho_1) + (1-\sigma)\kappa, \quad \gamma+\eta \leq \kappa,$$
$$\delta = (1-\sigma)(\kappa-(1-\rho_2)), \quad 0 \leq \sigma \leq 1,$$

we have

$$\delta-\eta+1-\varepsilon-(\gamma+\eta) = 1-\varepsilon-\sigma(1-\rho_1)-(1-\sigma)(1-\rho_2)-2\eta,$$
$$\geq \min(\rho_1,\rho_2)-\varepsilon-2\eta.$$

Assuming that $\kappa-(1-\rho_2) > 1-\rho_1$ we can start with $u(t) \in D(A^{1-\rho_1+\eta})$ for some $\eta > 0$ and reach the desired result after finitely many steps. If $\kappa-(1-\rho_2) \leq 1-\rho_1$ the result is proved by the following calculations; these are valid in both cases.

We have

$$A^{\delta-\eta+1-\varepsilon}u(t) = A^{1-\varepsilon}e^{-tA}A^{\delta-\eta}\varphi - \int_0^t A^{1-\varepsilon}e^{-(t-s)A}A^{\delta-\eta}M(u(s))\,ds.$$

This gives: $u(t) \in D(A^{\kappa-(1-\rho_2)+\min(\rho_1,\rho_2)-\varepsilon-2\eta})$, $0 < t \leq \tilde{T}$, $A^{\kappa-(1-\rho_2)+\min(\rho_1,\rho_2)-\varepsilon-2\eta}u(.) \in C^0((0,\tilde{T}],B)$. Considering the integral equation

$$u(t) = e^{-tA}u(t_0) - \int_{t_0}^{t_0+t} e^{-(t-s)A}M(u(s))\,ds$$

for some $t_0 \in (0,\tilde{T}]$ and for $t \in [0,\tilde{T}-t_0]$ we arrive at $A^{\kappa-(1-\rho_2)+2(\min(\rho_1,\rho_2)-\varepsilon-2\eta)}u(.) \in C^0((0,\tilde{T}],B)$ and so on. Finally we have $A^{\kappa}u(.) \in C^0((0,\tilde{T}],B)$ and

$$A^{\kappa-(1-\rho_2)+1-\varepsilon}u(t) = A^{1-\varepsilon}e^{-tA}A^{\kappa-(1-\rho_2)}\varphi$$
$$- \int_0^t A^{1-\varepsilon}e^{-(t-s)A}A^{\kappa-(1-\rho_2)}M(u(s))\,ds$$

thus getting the desired result. □

If we want to weaken the assumptions on the initial value we need additional conditions on the nonlinearity; these are growth conditions on $\|M(u)\|$. We assume that M fulfills the assumptions of theorem II.3.1. Moreover let

$$(\text{II.3.5})\begin{cases} \|M(u)\| \leq c(\|A^{1-\rho_1}u\|\,\|u\| + \|A^{1-\rho_1}u\|), \\ \|A^{\kappa-(1-\rho_2)}M(u)\| \leq c(\|A^{\kappa}u\|^{1+\rho_4}\|u\| + \|A^{\kappa}u\|) \end{cases}$$

with a constant $c>0$ and some $\rho_4 \in [0,1]$. Here u is from $D(A^{1-\rho_1})$, $D(A^{\kappa})$ respectively. The proof of theorem II.2.1, particularly the estimate of $\|A^{-\eta}\phi(\sigma)\|$, together with (II.3.5) furnishes the inequality

$$\|A^{-\eta}\phi(\sigma)\| \leq c(\eta)(\|A^{\eta}x\|^{(1+\rho_4)(1-\sigma)}\|A^{-\alpha_2+\eta}x\|^{1-\sigma} + \|A^{\eta}x\|^{1-\sigma})\cdot$$

$$\cdot(\|A^{\eta}x\|^{\sigma}\|A^{-\alpha_1+\eta}x\|^{\sigma} + \|A^{\eta}x\|^{\sigma});$$

thus if we set $x = A^{\gamma}u$ for some $u \in D(A^{\gamma+\eta})$ it follows that

$$\|A^{\delta-\eta}e^{-A\rho}M(e^{-A\varepsilon}u)\| \leq$$

$$\leq c(\eta)\|A^{\gamma+\eta}u\|\cdot(\|A^{\gamma+\eta-\alpha_1}u\|^{\sigma}+1)(\|A^{\gamma+\eta-\alpha_2}u\|^{1-\sigma}\|A^{\gamma+\eta}u\|^{\rho_4(1-\sigma)}+1),$$

where $\alpha_1 = 1-\rho_1$, $\beta_1 = 0$, $\alpha_2 = \kappa$, $\beta = \kappa-(1-\rho_2)$,

$$\gamma = (1-\sigma)\kappa + \sigma(1-\rho_1),$$

$$\delta = (1-\sigma)(\kappa-(1-\rho_2)), \quad \eta > 0.$$

Finally, for $\gamma+\eta \leq \alpha_1$, we get

$$(\text{II.3.6}) \quad \|A^{\delta-\eta}M(u)\| \leq c(\eta)\|A^{\gamma+\eta}u\|^{1+(1-\sigma)\rho_4}\cdot$$

$$\cdot(\|A^{\gamma+\eta-\alpha_1}u\|^{\sigma}+1)\|u\|^{1-\sigma} + c(\eta)\|A^{\gamma+\eta}u\|\cdot(\|A^{\gamma+\eta-\alpha_1}u\|^{\sigma}+1).$$

Now we prove a theorem which does not use the analyticity of M but only the first inequality in (II.3.5).

<u>Theorem II.3.3:</u> <u>Let</u> $\rho_1 \in (0,1)$. <u>Let</u> $M: D(A^{1-\rho_1}) \to B$ <u>be a Lipschitz continuous mapping fulfilling the estimate</u>

$$\|M(u)-M(v)\| \leq c[\|A^{1-\rho}(u-v)\| (\|u\|+\|v\|) + \\ + (\|A^{1-\rho}u\|+\|A^{1-\rho}v\|)\|u-v\|],$$

$u, v \in D(A^{1-\rho_1})$. <u>Let</u> $\varphi \in B$. <u>Then there exists a</u> $T_\varphi = T_\varphi(\varphi) \in (0,+\infty]$ <u>with the following property</u>: <u>There is one and only one mapping</u> $u:[0,T_\varphi) \to B$ <u>with</u>

$$u(.) \in \bigcap_{\tilde{T}, 0<\tilde{T}<T_\varphi} C^0([0,\tilde{T}],B),$$

$$u(t) \in D(A^{1-\rho_1}),\ 0<t<T_\varphi,\ \cdot^{1-\rho_1}A^{1-\rho_1}u(.) \in \bigcap_{\tilde{T}, 0<\tilde{T}<T(\varphi)} L^\infty((0,\tilde{T}),B)$$

$$A^{1-\rho_1}u(.) \in \bigcap_{\varepsilon, \tilde{T}, 0<\varepsilon<\tilde{T}<T_\varphi} C^0([\varepsilon,\tilde{T}],B),$$

$$u(t) = e^{-tA}\varphi - \int_0^t e^{-(t-s)A}M(u(s))\, ds,$$

$$\lim_{t\uparrow T_\varphi} \|u(t)\| = +\infty$$

<u>if</u> $T_\varphi < +\infty$.

<u>Proof:</u> Let us consider the complete metric space

$$\mathfrak{M}_T = \Big\{u \mid u:[0,\tilde{T}] \to B,\ u(0) = \varphi,\ u(t) \in D(A^{1-\rho_1}),\ 0<t\leq\tilde{T}, \\ u(.) \in C^0([0,\tilde{T}],B),\ A^{1-\rho_1}u(.) \in \bigcap_{\varepsilon, 0<\varepsilon<\tilde{T}} C^0([\varepsilon,\tilde{T}],B), \\ \|t^{1-\rho_1}A^{1-\rho_1}u(.)\| \in L^\infty((0,\tilde{T}),B)\Big\}$$

endowed with the metric

$$\mu_{\tilde{T}}(u_1,u_2) = \sup_{0\le t\le\tilde{T}} \|u_1(t)-u_2(t)\| + \sup_{0<t\le\tilde{T}} t^{1-\rho_1}\|A^{1-\rho_1}(u_1(t)-u_2(t))\|.$$

Let $v \in \mathfrak{M}_{\tilde{T}}$. Then

$$\mathcal{T}v(t) = e^{-tA}\varphi - \int_0^t e^{-(t-s)A}M(v(s))\, ds$$

defines a mapping from $\mathfrak{M}_{\tilde{T}}$ into $C^0([0,\tilde{T}],B)$; for the integral $\int_0^t e^{-(t-s)A}M(v(s))\, ds$ the following inequalities hold:

$$\|\int_0^t A^{1-\rho_1}e^{-(t-s)A}M(v(s))\, ds\|$$

$$\le \int_0^t \frac{c}{(t-s)^{1-\rho_1}}\|A^{1-\rho_1}v(s)\|\cdot\|v(s)\|\, ds,$$

$$\le \int_0^t \frac{c}{(t-s)^{1-\rho_1}s^{1-\rho_1}}\, ds \sup_{0<t\le\tilde{T}} t^{1-\rho_1}\|A^{1-\rho_1}v(t)\| \sup_{0\le t\le\tilde{T}}\|v(\tilde{t})\|\, ds,$$

$$\|t^{1-\rho_1}A^{1-\rho_1}\mathcal{T}v(t)\| \le c\|\varphi\| + ct^{\rho_1} \sup_{0<t\le\tilde{T}} t^{1-\rho_1}\|A^{1-\rho_1}v(t)\|\cdot$$

$$\cdot \sup_{0\le t\le\tilde{T}}\|v(t)\|.$$

Thus we see that $\mathcal{T}v(t) \in D(A^{1-\rho_1})$, $0 < t \le \tilde{T}$, $\mathcal{T}v(0) = \varphi$. Because of the growth condition on $M(u)$ we get

$$s^{1-\rho_1}M(v(s)) \in L^\infty((0,\tilde{T}),B).$$

Because of the strong continuity of $A^{1-\rho_1}e^{-tA}$ and the estimate (I.3.4) it is not difficult to show that the expression $\int_0^t A^{1-\rho_1}e^{-(t-s)A}M(u(s))\, ds$ is in $C^0((0,\tilde{T}],B)$. It follows that

$Tv \in \mathfrak{M}_{\tilde{T}}$. For $Tv_1(t)-Tv_2(t)$, $v_1,v_2 \in \mathfrak{M}_{\tilde{T}}$, we get

$$\|t^{1-\rho_1}A^{1-\rho_1}(Tv_1(t)-Tv_2(t))\|$$

$$\leq \int_0^t \frac{ct^{1-\rho_1}}{(t-s)^{1-\rho_1}s^{1-\rho_1}}\,ds \cdot \sup_{0<t\leq\tilde{T}} t^{1-\rho_1}\|M(v_1(\tilde{t}))-M(v_2(\tilde{t}))\|,$$

$$\leq ct^{\rho_1}\Big[\sup_{0<t\leq\tilde{T}} \tilde{t}^{1-\rho_1}\|A^{1-\rho_1}v_1(\tilde{t})-A^{1-\rho_1}v_2(\tilde{t})\|\cdot \sup_{0\leq t\leq\tilde{T}}(\|v_1(\tilde{t})\|+\|v_2(\tilde{t})\|$$

$$+ \sup_{0<t\leq\tilde{T}}(\tilde{t}^{1-\rho_1}\|A^{1-\rho_1}v_1(\tilde{t})\|+\tilde{t}^{1-\rho_1}\|A^{1-\rho_1}v_2(\tilde{t})\|)\cdot$$

$$\cdot \sup_{0<t\leq\tilde{T}}\|v_1(\tilde{t})-v_2(\tilde{t})\|\Big].$$

An analogous estimate can be proved for $\|Tv_1(t)-Tv_2(t)\|$. It is easily seen that for a sufficient small $\tilde{T}$ the mapping T is a contraction if T is restricted to a set $\mu_{\tilde{T}}(u,0) \leq 2C$. Here $C>0$ and

$$\sup_{0\leq t\leq\tilde{T}} \|e^{-\tilde{t}A}\varphi\| \leq C,$$

$$\sup_{0<t\leq\tilde{T}} t^{1-\rho_1}\|A^{1-\rho_1}e^{-tA}\varphi\| \leq C.$$

Banach's fixed point theorem now furnishes the desired result with the exception of the "explosion property".

For the fixed point u of T we get the estimate

$$\|A^{1-\rho_1}u(t)\| \leq \frac{c}{t^{1-\rho_1}}\|\varphi\| + \int_0^t \frac{c}{(t-s)^{1-\rho_1}}\|A^{1-\rho_1}u(s)\|\,\|u(s)\|\,ds.$$

Thus $\|A^{1-\rho_1}u(t)\| \leq \frac{c}{t^{1-\rho_1}}\, g(\sup_{0\leq t\leq\tilde{T}}\|u(t)\|)$, $0<t\leq\tilde{T}$ (cf. the proof of (I.2.2)) and $\|A^{1-\rho_1}u(t)\|$ can be estimated in terms of $\|u(t)\|$; $g:[0,+\infty)\to[0,+\infty)$ is continous. This gives the desired result. □

The following theorem is a slightly strengthened version of theorem II.3.3.

Theorem II.3.4: Let ρ_o, ρ_1 be from $(0,1)$ and let $(1-\rho_1)(1+\rho_o) = 1-\rho_1' < 1$ with some $\rho_1' \in (0,1)$. We assume that

$$M: D(A^{1-\rho_1}) \to B$$

is a Lipschitz continuous mapping satisfying the estimate

$$\|M(u)-M(v)\| \leq c\Big[\|A^{1-\rho_1}(u-v)\| \cdot (\|A^{1-\rho_1}u\|^{\rho_o} + \|A^{1-\rho_1}v\|^{\rho_o}) \cdot$$
$$(\|u\| + \|v\| + 1) + (\|A^{1-\rho_1}u\|^{1+\rho_o} + \|A^{1-\rho_1}v\|^{1+\rho_o} + 1)\ \|u-v\|\Big].$$

Then the statement of theorem II.3.3 remains true without the last two lines. Instead of it we have

$$\lim_{t \uparrow T_\varphi} \int_\eta^t \|A^{1-\rho_1}u(s)\|^2\, ds = +\infty, \quad 0 < \eta < T_\varphi,$$

if $T_\varphi < +\infty$.

Proof: The proof can be given as the proof of theorem II.3.3 - with the exception of the explosion property. One has to substitute the factor $s^{-(1-\rho_1)(1+\rho_o)}$ $s^{(1-\rho_1)(1+\rho_o)}$ in the integral only. Observe that $(1-\rho_1)(1+\rho_o) = 1-\rho_1' < 1$ according to our assumption.

As for the "explosion property" we have

$$\|A^{1-\rho_1}u(t)\| \leq c\|A^{1-\rho_1}u(\eta)\| + \int_\eta^t \frac{c}{|t-s|^{1-\rho_1}} \cdot$$
$$\cdot \|A^{1-\rho_1}u(s)\|^{1+\rho_o}\|u(s)\|\ ds$$

for some $\tilde{\eta} \in (0,\tilde{T})$ and any $t \in [\tilde{\eta},\tilde{T}]$, where $[0,\tilde{T}]$ is an existence interval. Since $(1-\rho_1)(1+\rho_0) < 1$ we can use Kielhöfer's lemma with $= \frac{3-\rho_0}{2(3+\rho_0)}$ and bound $\sup_{\eta \le t \le \tilde{T}} \|A^{1-\rho_1} u(s)\|$ in terms of $\int_{\eta}^{\tilde{T}} \|A^{1-\rho_1} u(s)\|^2 ds$ (For Kielhöfer's lemma see the comments to this paragraph). Thus our theorem is proved. □

Let us remark that the linear growth with respect to u in the Lipschitz condition of theorem II.3.4 could be replaced by any expression $g(\|u\|)$ with a continuous function g. The higher regularity of the solutions constructed in the preceding theorems can be considered with the aid of the analyticity of the mapping M.

Theorem II.3.5: *We assume that the nonlinearity* M *fulfills the Lipschitz estimate of theorem* II.3.3, *the assumptions of theorem* II.3.1 *and the growth condition* (II.3.5). *Let* $\varphi \in B$. *Then the solution of the integral equation* $u(t) = e^{-tA}\varphi + \int_0^t e^{-(t-s)A} M(u(s))\, ds$, *which has been constructed in theorem* II.3.3, *has the following additional regularity properties:*

$$u(t) \in D(A^{\kappa+\rho_2-\hat{\eta}}), \quad 0 < t < T, \quad 0 < \hat{\eta},$$

$$A^{\kappa+\rho_2-\hat{\eta}} u(.) \in C^0((0,T_\varphi),B).$$

T_φ *is the quantity having been introduced in theorem* II.3.3. *Moreover* κ *must fulfill the inequality*

$$\kappa < 1+\frac{1}{1+\rho_4}.$$

Proof: For ε, $0 < \varepsilon \le 1$, we have

$$A^{1-\varepsilon}u(t) = A^{1-\varepsilon}e^{-tA}\varphi - \int_0^t A^{1-\varepsilon}e^{-(t-s)A}M(u(s))\, ds,$$

where

$$\|M(u(s))\| \leq \sup_{0\leq s\leq t} (\|u(s)\|+1)\cdot \frac{c}{s^{1-\rho_1}},$$

$$\|A^{1-\varepsilon}e^{-(t-s)A}\| \leq \frac{c}{(t-s)^{1-\varepsilon}},\ 0 \leq s < t.$$

Since $A^{1-\varepsilon}e^{-(t-s)A}x$, $x \in B$, is continuous, the expression $A^{1-\varepsilon}e^{-tA}\varphi$ is continuous in t, $t > 0$. Moreover, because of our Lipschitz condition on M, we have

$$\cdot^{1-\rho_1} M(u(.)) \in L^{\infty}((0,\tilde{T}),B),\ \tilde{T} < T_{\varphi}.$$

Therefore with (II.3.3) it follows that $\int_0^{\cdot} A^{1-\varepsilon}e^{-(.-s)A}M(u(s))\,ds$ is continuous on $(0,T_{\varphi})$. Now let $\tilde{\eta} \in (0,T_{\varphi})$, $\tilde{\eta} < \tilde{T}$. Then

$$u(t) = e^{-(t-\tilde{\eta})A}u(\tilde{\eta}) - \int_{\tilde{\eta}}^{t} e^{-(t-s)A}M(u(s)),\ \tilde{\eta} \leq t < T_{\varphi}.$$

We set

$$\delta = \delta(\sigma) = (1-\sigma)(\kappa-(1-\rho_2)),$$

$$\gamma = \gamma(\sigma) = (1-\sigma)\kappa + \sigma(1-\rho_1).$$

Then we get

$$A^{\delta-\eta+1-\varepsilon}u(t) =$$

$$= A^{\delta-\eta+1-\varepsilon}e^{-(t-\tilde{\eta})A}u(\tilde{\eta}) - \int_{\tilde{\eta}}^{t} A^{\delta-\eta+1-\varepsilon}e^{-(t-s)A}M(u(s))\,ds,$$

$$= A^{\delta-\eta}e^{-(t-\tilde{\eta})A}A^{1-\varepsilon}u(\tilde{\eta}) - \int_{\tilde{\eta}}^{t} A^{1-\varepsilon}e^{-(t-s)A}A^{\delta-\eta}M(u(s))\,ds,$$

where, according to (II.3.6)

$$\|A^{\delta-\eta}M(u(s))\| \leq c(\eta)\|A^{\gamma+\eta}u(s)\|^{1+(1-\sigma)\rho_4}\cdot$$

$$\cdot(\|A^{\gamma+\eta-(1-\rho_1)}u(s)\|^{\sigma}+1)\|u(\sigma)\|^{1-\sigma} +$$

$$+ c(\eta)\|A^{\gamma+\eta}u\|\cdot(\|A^{\gamma+\eta-\alpha_1}u\|^{\beta}+1).$$

Let $\gamma+\eta \geq 1-\varepsilon \geq 1-\rho_1$, $A^{\gamma+\eta}u(.) \in C^o((\tilde{\eta},T_\varphi),B)$, and

$$\|A^{\gamma+\eta}u(s)\| \leq \frac{c}{(s-\tilde{\eta})^{\gamma+\eta-(1-\varepsilon)}},$$

$$\|A^{\gamma+\eta-(1-\rho_1)}u(s)\| \leq \frac{c}{(s-\tilde{\eta})^{\max(\gamma+\eta-(1-\rho_1)-(1-\varepsilon),0)}}.$$

Then we have $(1+(1-\sigma)\rho_4)(\gamma+\eta-(1-\varepsilon)) + \sigma\cdot\max(\gamma+\eta-(1-\rho_1)-(1-\varepsilon),0) = (1+(1-\sigma)\rho_4)(\gamma+\eta-(1-\varepsilon))$, if $\gamma+\eta \leq 1-\rho_1+1-\varepsilon$. From our assumption we get that $\gamma \leq \kappa < 2$. If (II.3.5) holds for some $\rho_1 \in (0,1)$, then it does for all ρ_1', $0<\rho_1' \leq \rho_1$. Thus we can always assume that

$$\gamma+\eta \leq 1-\rho_1+1-\varepsilon,$$

if η,ε are sufficiently small. The last integral term behaves like

$$\int_{\tilde{\eta}}^{t} \frac{c}{(t-s)^{1-\varepsilon}}\cdot\frac{1}{(s-\tilde{\eta})^{(\gamma+\eta-(1-\varepsilon))(1+(1-\sigma)\rho_4)}}\,ds,$$

i.e. like

$$\frac{1}{(t-\tilde{\eta})^{1-\varepsilon+(\gamma+\eta-(1-\varepsilon))(1+(1-\sigma)\rho_4)-1}};$$

the same argument as above shows that it is continuous in $(\tilde{\eta},T_\varphi)$. Now we consider the functions

$$L(\sigma) = \delta(\sigma)+1 = (1-\sigma)(\kappa-(1-\rho_2))+1,$$

$$R(\sigma) = \gamma(\sigma)(1+(1-\sigma)\rho_4) - (1-\varepsilon)(1+(1-\sigma)\rho_4).$$

We have $L(\sigma) > R(\sigma)$,

$$0 \leq \varepsilon \leq \inf_{0\leq\sigma\leq 1} \frac{(1-\sigma)(\kappa-(1-\rho_2))+1.(\frac{1-\sigma}{1+\rho_4}-\sigma\rho_1)(1+(1-\sigma)\rho_4}{1+(1-\sigma)\rho_4} \quad \text{if } \gamma<\kappa<$$

$1+\frac{1}{1+\rho_4}$. For sufficiently small $\varepsilon,\eta>0$ it therefore follows that

$$\delta-\eta > 1-\varepsilon+(\gamma+\eta-(1-\varepsilon))(1+(1-\sigma)\rho_4)-1,$$

$$\delta-\eta-[(1-\varepsilon)+(\gamma+\eta-(1-\varepsilon))(1+(1-\sigma)\rho_4)-1] \geq \varepsilon_o > 0,\ 0\leq\sigma\leq 1.$$

Also for sufficiently small $\varepsilon,\eta > 0$ we have $1-\varepsilon_o \geq R(\sigma)$, $0 \leq \sigma \leq 1$. On $[0,1]$ we always have $\delta+1 > \gamma$. Thus we can assume that

$$\delta-\eta+1-\varepsilon-(\gamma+\eta) \geq \varepsilon_o > 0, \quad 0 \leq \sigma \leq 1,$$

if ε,η are sufficiently small but fixed. There is exactly one $\sigma_1 \in [0,1]$ with

$$\gamma_1 = \gamma(\sigma_1) = 1-\varepsilon-\eta.$$

The corresponding $\delta_1 = \delta(\sigma_1)$ fulfills the following relations:

$$\delta_1-\eta+1-\varepsilon-(\gamma_1+\eta) \geq \varepsilon_o,$$

$$u(t) \in D(A^{\delta_1-\eta+1-\varepsilon}), \quad \tilde{\eta} < t < T_\varphi,$$

$$A^{\delta_1-\eta+1-\varepsilon}u(.) \in C^o((\tilde{\eta},T_\varphi),B),$$

$$\|A^{\delta_1-\eta+1-\varepsilon}u(t)\| \leq \frac{c}{(t-\tilde{\eta})^{\delta_1-\eta}}, \quad \tilde{\eta} < t < T_\varphi.$$

Let $\hat{\eta} \in (0,1)$ be fixed. If already $\delta_1-\eta+1-\varepsilon = \kappa-(1-\rho_2)+1-\hat{\eta}$ there is nothing to prove. It remains to consider the case $\delta_1-\eta+1-\varepsilon < \kappa$. Then there is again exactly one $\sigma_2 \in [0,1]$ with $\sigma_2 < \sigma_1$,

$$\gamma(\sigma_2) = \delta_1-\eta+1-\varepsilon-\eta.$$

We set $\delta_2 = \delta(\sigma_2)$. If $\delta_2-\eta+1-\varepsilon < \kappa$ we continue again, and so on. After finitely many steps we arrive at the desired result. □

Now we prove an analogue to theorem II.3.3 requiring weaker growth conditions on the nonlinearity; in exchange of this we assume that the initial value φ is in $D(A^\delta)$ for some $\delta \geq 0$.

<u>Theorem II.3.6:</u> <u>Let</u> M <u>be a mapping from</u> $D(A^{1-\rho_1})$ <u>into</u> B <u>with</u>

$$\|M(u)\| \leq c\|A^{1-\rho_1}u\|^{1+\rho_o}(\|A^\delta u\|+1),$$

$u \in D(A^{1-\rho_1})$. ρ_o <u>is a number</u> ≥ 0, ρ_1 <u>is from</u> $(0,1)$. δ <u>is</u> > 0. <u>We</u>

<u>assume that</u>

$$0 < 1-\rho_1 - \frac{\rho_1}{\rho_0} \leq \delta < 1-\rho_1 \ \underline{\text{if}}\ \rho_0 > 0$$

<u>and</u>

$$\delta \leq 1-\rho_1 \ \underline{\text{if}}\ \rho_0 = 0.$$

<u>Moreover</u>, M <u>fulfills the following Lipschitz condition</u>:

$$\begin{aligned} \|M(u)-M(v)\| \leq\ & c(\|A^\delta u\|+\|A^\delta v\|+1)(\|A^{1-\rho_1}u\|^{\rho_0} + \\ & + \|A^{1-\rho_1}v\|^{\rho_0}+1)\|A^{1-\rho_1}(u-v)\| + \\ & + c(\|A^{1-\rho_1}u\|+\|A^{1-\rho_1}v\|+1)\cdot\|A^\delta(u-v)\|^{1-\rho_0}\cdot \\ & \cdot \|A^{1-\rho_1}(u-v)\|^{\rho_0}. \end{aligned}$$

<u>Let</u> $\varphi \in D(A^\delta)$. <u>Then there exists a</u> $T_\varphi \in (0,+\infty]$ <u>with the following properties</u>: <u>There is exactly one</u> $u:[0,T_\varphi) \to B$ <u>with</u>

$$u \in C^0([0,T_\varphi),B),$$

$$u(t) \in D(A^{1-\rho_1}),\ 0 < t < T_\varphi$$

$$\cdot^{1-\rho_1-\delta}\|A^{1-\rho_1}u(.)\| \in \bigcap_{0<\tilde{T}<T(\varphi)} L^\infty((0,\tilde{T}))$$

$$A^{1-\rho_1}u(.) \in C^0((0,T_\varphi),B),$$

$$A^\delta u(.) \in \bigcap_{0<\tilde{T}<T_\varphi} L^\infty((0,\tilde{T}),B),$$

$$u(t) = e^{-tA}\varphi - \int_0^t e^{-(t-s)A}M(u(s))\,ds,$$

$$\lim_{t\uparrow T_\varphi} \|A^{1-\rho_1}u(t)\| = \infty \ \underline{\text{if}}\ T_\varphi < +\infty.$$

<u>In particular we have</u>

$$A^{1-\rho_1}u(.) \in \bigcap_{0<\tilde{T}<T_\varphi} L^{1+\rho_0}((0,\tilde{T}),B).$$

Proof: Let $\rho_o > 0$. We want to apply Banach's fixed point theorem for small $\tilde{T}$. For any $\tilde{T} > 0$ we consider the space

$$\mathfrak{M}_{\tilde{T}} = \Big\{ v \Big| v \in C^o([0,\tilde{T}], D(A^{\delta})),\ v(0) = \varphi,\ v(t) \in D(A^{1-\rho_1}),\ 0 < t \leq \tilde{T},$$

$$\cdot^{1-\rho_1-\delta} A^{1-\rho_1} v(.) \in \bigcap_{\varepsilon, 0<\varepsilon<\tilde{T}} C^o([\varepsilon,\tilde{T}],B) \cap L^{\infty}((0,\tilde{T}),B),$$

$$\omega_v(t) = \| t^{1-\rho_1-\delta} A^{1-\rho_1} e^{-tA} v(t) \| \leq 2 \sup_{0<t\leq\tilde{T}} \| t^{1-\rho_1-\delta} A^{1-\rho_1} e^{-tA} \varphi \|,$$

$$0 < t \leq \tilde{T},\ \sup_{0\leq t\leq\tilde{T}} \| A^{\delta} v(t) \| \leq \sup_{0\leq t\leq\tilde{T}} \| A^{\delta} e^{-tA} \varphi \| + 1 \Big\}.$$

Moreover we set for $v_1, v_2 \in \mathfrak{M}_{\tilde{T}}$

$$\mu_{\tilde{T}}(v_1,v_2) = \operatorname*{ess\,sup}_{t\in(0,\tilde{T})} \| A^{\delta}(v_1-v_2)(t) \| + \operatorname*{ess\,sup}_{t\in(0,\tilde{T})} t^{1-\rho_1-\delta} \| A^{1-\rho_1}(v_1-v_2)(t) \|.$$

Endowed by $\mu_{\tilde{T}}$ as a metric, $\mathfrak{M}_{\tilde{T}}$ is a complete metric space (Observe that for $\delta \geq 0$ the inequality $\|x\| \leq c\|A^{\delta}x\|$ holds). First of all we remark that

$$\| t^{1-\rho_1-\delta} A^{1-\rho_1} e^{-tA} \varphi \| \to 0 \text{ for } t \to 0.$$

We define a mapping T by

$$Tv(t) = e^{-tA}\varphi - \int_0^t e^{-(t-s)A} M(v(s))\, ds.$$

First we want to show that T is a mapping from $\mathfrak{M}_{\tilde{T}}$ in itself if $\tilde{T}$ is sufficiently small. Since

$$\begin{aligned}(1-\rho_1-\delta)(1+\rho_o) &\leq \frac{\rho_1}{\rho_o}(1+\rho_o), \\ &\leq \frac{\rho_1}{\rho_o} + \rho_1, \\ &\leq 1-\delta,\end{aligned}$$

we get: $A^{1-\rho_1} v(.) \in L^{1+\rho_o}((0,\tilde{T}),B)$. Thus $M(v(.)) \in L^1((0,\tilde{T}),B)$ and $\int_0^t e^{-(t-s)A} M(v(s))\, ds$ is continuous on $[0,\tilde{T}]$. Evidently $\mathcal{T}v(0) = \varphi$, $\cdot^{1-\rho_1-\delta} A^{1-\rho_1} e^{-\cdot A}\varphi \in L^\infty((0,\tilde{T}),B) \cap C^0((0,\tilde{T}],B)$. Also

$$t^{1-\rho_1-\delta} \int_0^t A^{1-\rho_1} e^{-(t-s)A} M(v(s))\, ds =$$

$$= t^{1-\rho_1-\delta} \int_0^t A^{1-\rho_1} e^{-(t-s)A} \frac{s^{(1-\rho_1-\delta)(1+\rho_o)}}{s^{(1-\rho_1-\delta)(1+\rho_o)}} M(v(s))\, ds$$

$\in L^\infty((0,\tilde{T}),B)$. The continuity of $\int_0^t A^{1-\rho_1} e^{-(t-s)A} M(v(s))\, ds$ follows as in the proof of theorem II.3.3. Moreover

$$\| t^{1-\rho_1-\delta} \int_0^t A^{1-\rho_1} e^{-(t-s)A} M(v(s))\, ds \| \le$$

$$\le c \cdot \sup_{0<s\le\tilde{T}} \| s^{1-\rho_1-\delta} A^{1-\rho_1} v(s) \|^{1+\rho_o} \cdot (\sup_{0<s\le\tilde{T}} \| A^\delta v(s) \| + 1),$$

(II.3.7)

$$\le c \cdot \sup_{0<s\le\tilde{T}} \| s^{1-\rho_1-\delta} A^{1-\rho_1} e^{-sA}\varphi \|^{1+\rho_o} \cdot (\sup_{0<s\le\tilde{T}} \| A^\delta v(s) \| + 1),$$

$$\| A^\delta \int_0^t e^{-(t-s)A} M(v(s))\, ds \| \le c \int_0^t \frac{ds}{(t-s)^\delta s^{(1-\rho_1-\delta)(1+\rho_o)}} \cdot$$

$$\cdot \sup_{0<s\le\tilde{T}} \| s^{1-\rho_1-\delta} A^{1-\rho_1} v(s) \|^{1+\rho_o} \cdot (\sup_{0<s\le\tilde{T}} \| A^\delta v(s) \| + 1),$$

$$\le c \cdot \sup_{0<s\le\tilde{T}} \| s^{1-\rho_1-\delta} A^{1-\rho_1} e^{-sA}\varphi \|^{1+\rho_o} \cdot (\sup_{0<s\le\tilde{T}} \| A^\delta v(s) \| + 1).$$

Because of $\rho_o > 0$ it in particular follows from (II.3.7) that $\mathcal{T}$ is a mapping from $\mathfrak{M}_{\tilde{T}}$ into itself if $\tilde{T}$ is sufficiently small. It is easy to see that $\mathcal{T}$ is also a contraction ($\tilde{T}$ sufficiently small)

The smallness of the coefficients being necessary in the Lipschitz estimate for $\|t^{1-\rho_1-\delta} A^{1-\rho_1} (\int_0^t e^{-(t-s)A} M(v_1(s))\, ds - \int_0^t e^{-(t-s)A} M(v_2(s))\, ds\|$ is mainly provided by the estimate for $\omega_{v_1}(t)$, $\omega_{v_2}(t)$ and the corresponding factors in the Lipschitz estimate for $\|M(v_1(s))-M(v_2(s))\|$ (Observe that $\rho_o > 0$). Thus our theorem is proved in the case $\rho_o > 0$. For the case $\rho_o = 0$ confer theorem II.3.3. □

Let us remark that from our proof it follows that

(II.3.8) $A^{\delta}u \in C^{o}([0,\tilde{T}],B)$,

in theorem II.3.6.

Next we want to give a new version of theorem II.3.5. Here we only need a weeker version of the condition (II.3.5) but on the other hand we must assume that a solution of the equation

$$u(t) = e^{-tA}\varphi - \int_0^t e^{-(t-s)A} M(u(s))$$

already exists. Let $M \in A(D(A^{1-\rho_1}),B)$ and let

$$\|M(u)\| \le c\|A^{1-\rho_1}u\|^{1+\rho_o}(\|u\|+1), \quad u \in D(A^{1-\rho_1}).$$

Moreover we assume that $M \in A(D(A^{\kappa}), D(A^{\kappa-(1-\rho_2)})$ and

$$\|A^{\kappa-(1-\rho_2)} M(u)\| \le c\|A^{\kappa} u\|^{1+\rho_4}(\|u\|+1).$$

Here ρ_1, ρ_2 are in $(0,1)$ whereas $\rho_o, \rho_4 \ge 0$. As before we get

$$\|A^{-\eta}\phi(\sigma)\| \le c(\eta)\|A^{\eta}x\|^{(1-\sigma)(1+\rho_4)}(\|A^{-\alpha_2+\eta}x\|^{1-\sigma} + 1)\cdot$$

$$\cdot\|A^{\eta}x\|^{\sigma(1+\rho_o)}(\|A^{-\alpha_1+\eta}x\|^{\sigma} + 1).$$

From this it follows with $u \in D(A^{\gamma+\eta})$, $x = A^{\gamma}u$, $\gamma+\eta \leq \alpha_2$ that

$$\|A^{\delta-\eta}M(u)\| \leq c(\eta)\|A^{\gamma+\eta}u\|^{1+(1-\sigma)\rho_4+\sigma\rho_0} \cdot (\|u\|^{1-\sigma}+1)(\|A^{\gamma+\eta-(1-\rho_1)}u\|^{\sigma}+1);$$

here $\alpha_1 = 1-\rho_1$, $\beta_1 = 0$, $\alpha_2 = \kappa$, $\beta_2 = \kappa-(1-\rho_2)$,

$$\gamma = \gamma(\sigma) = (1-\sigma)\kappa + \sigma(1-\rho_1),$$
$$\delta = \delta(\sigma) = (1-\sigma)(\kappa-(1-\rho_2)).$$

Using the inequality for $\|A^{\delta-\eta}M(u)\|$ we are now able to prove

<u>Theorem II.3.7:</u> <u>Let</u> $\kappa \leq 1$, <u>let</u> $\varphi \in B$, <u>let</u> $T_0 > 0$. <u>Let</u>

$$u \in C^0([0,T_0],B),$$
$$u(t) \in D(A^{1-\rho_1}),\ 0 < t \leq T_0,$$
$$A^{1-\rho_1}u(.) \in C^0((0,T_0],B) \cap L^{1+\rho_0}((0,T_0),B),$$
$$u(t) = e^{-tA}\varphi - \int_0^t e^{-(t-s)A}M(u(s))\,ds,\ 0 \leq t \leq T_0.$$

<u>Then</u>

$$u(t) \in \bigcap_{0<\hat{\eta}\leq 1} D(A^{\kappa-(1-\rho_2)+1-\hat{\eta}}),\ 0 < t \leq T_0,$$
$$A^{\kappa-(1-\rho_2)+1-\hat{\eta}}u(.) \in C^0((0,T_0],B),\ 0 < \hat{\eta} \leq 1,$$
$$u \in C^1((0,T_0],B),$$
$$u' + Au + M(u) = 0,\ 0 < t \leq T_0,$$
$$u(0) = \varphi.$$

Proof: We assume as in the proof of theorem II.3.5 that $\tilde{\eta} > 0$ and consider the equation

$$\text{(II.3.9)} \quad u(t) = e^{-(t-\tilde{\eta})A}u(\tilde{\eta}) - \int_{\tilde{\eta}}^{t} e^{-(t-s)A}M(u(s))\, ds.$$

Let us assume that

$$\text{(II.3.10)} \quad \begin{array}{l} u(t) \in \bigcap_{0<\varepsilon\leq 1} D(A^{1-\varepsilon}),\ 0 < t \leq T_o, \\ A^{1-\varepsilon}u(.) \in C^o((0,T_o],B),\ 0 < \varepsilon \leq 1. \end{array}$$

In what follows this assumption will be justified. Let $\kappa = 1$, i.e. $\gamma(\sigma) = 1-\sigma+\sigma(1-\rho_1)$. Let ρ' be a number with $0 < \rho' < \kappa-(1-\rho_2)$. For sufficiently small $\sigma_1, \eta > 0$ we get a

$$\gamma' = \gamma(\sigma_1)+\eta$$

with $0 < \gamma' < 1$ such that $\rho' = \delta(\sigma_1)-\eta$, $M:D(A^{\gamma'}) \to D(A^{\rho'})$,

$$\|A^{\rho'}M(u)\| \leq c\|A^{\gamma'}u\|^{1+(1-\sigma_1)\rho_4+\sigma_1\rho_o} \cdot (\|u\|^{1-\sigma_1}+1)(\|A^{\gamma'-(1-\rho_1)}u\|^{\sigma_1}+1).$$

We have

$$\|A^{1-\varepsilon+\rho'}e^{-(t-s)A}x\| \leq \frac{c}{(t-s)^{1-\varepsilon}}\|A^{\rho'}x\|,$$

$x \in D(A^{\rho'})$, and $A^{1-\varepsilon+\rho'}e^{-(t-s)A}x$ is continuous on $t > s$. Thus it follows that

$$\int_{\tilde{\eta}}^{t} e^{-(t-s)A}M(u(s))\, ds \in D(A^{1-\varepsilon+\rho'}),$$

$$A^{1-\varepsilon+\rho'} \int_{\tilde{\eta}}^{\cdot} e^{-(\cdot-s)A}M(u(s))\, ds \in C^o([\tilde{\eta},T_o],B).$$

Since $\varepsilon, \tilde{\eta} > 0$ and $\rho' \in (0,\kappa-(1-\rho_2))$ were arbitrary the proof is completed if we can show (II.3.10). But this follows from the integral equation (II.3.9), the continuity of $A^{1-\varepsilon}e^{-(t-s)A}x$ on

$t > s$ and the fact that according to our assumptions $M(u(.)) \in$ $\in C^0((0,T_0],B)$. Evidently the case $\kappa < 1$ is the trivial one. □

As it is well known the solution of the integral equation

$$u(t) = e^{-tA}\varphi - \int_0^t e^{-(t-s)A} M(u(s))\, ds$$

is often interpreted as a weak solution of the differential equation

$$u' + Au + M(u) = 0, \quad u(0) = \varphi.$$

We now want to introduce a weaker notion of solution by multiplying the integral equation or the differential equation by $A^{-\delta}$. This will be done in the next theorem.

Theorem II.3.8: *For some* $\rho_1 \in [0,1)$ *let* M *be a mapping from* $D(A^{1-\rho_1})$ *into* B. *For some* $\delta \in (0,1)$ *and for* $u,v \in D(A^{1-\rho_1+\delta})$ *we set*

$$\tilde{M}(u) = M(A^{\delta}u)$$

and we assume that the following estimates are fulfilled:

$$\|A^{-\delta}\tilde{M}(u)\| \le c\|A^{1-\rho_1'}u\|^2(\|A^{\delta}u\| + 1),$$

$$\|A^{-\delta}(\tilde{M}(u)-\tilde{M}(v))\| \le c(\|A^{\delta}u\| + \|A^{\delta}v\| + 1)(\|A^{1-\rho_1'}u\| + \|A^{1-\rho_1'}v\|)\cdot$$
$$\cdot\|A^{1-\rho_1'}(u-v)\| +$$
$$+ c\cdot(\|A^{1-\rho_1'}u\|^2 + \|A^{1-\rho_1'}v\|^2)\cdot\|A^{\delta}(u-v)\|.$$

Here ρ_1' *is some number from* $(0,1)$ *with*

$$0 < 1-2\rho_1' \le \delta < 1-\rho_1'.$$

Then there exists for each $\varphi \in B$ *a* $T_\varphi \in (0,+\infty]$ *with the following property: There is one and only one* $u:[0,T_\varphi) \to B$ *with*

$u(.) \in C^{O}([O,T_{\varphi}),B),$

$$u(t) \in \bigcap_{O<\varepsilon'<1} D(A^{1-\varepsilon'-\delta}),$$

$$A^{1-\varepsilon'-\delta}u(.) \in C^{O}((O,T_{\varphi}),B),$$

$$\cdot^{1-\varepsilon'-\delta}A^{1-\varepsilon'-\delta}u(.) \in \bigcap_{O<T<T_{\varphi}} L^{\infty}((O,T),B),\ O<\varepsilon'<1,$$

$$A^{-\delta}u(t) = e^{-tA}A^{-\delta}\varphi - \int_{O}^{t} e^{-(t-s)A}A^{-\delta}M(A^{\delta}A^{-\delta}u(s))\,ds,\ O \leqq t < T_{\varphi}.$$

_If_ $T_{\varphi} < +\infty$ _then we have_

$$\lim_{t \uparrow T_{\varphi}} \|u(t)\| = \lim_{t \uparrow T_{\varphi}} \|A^{\delta}A^{-\delta}u(t)\| = +\infty \ \underline{\text{or}}$$

(II.3.11)

$$\omega(\tilde{\delta},T) = \sup_{O \leqq t \leqq \tilde{\delta},\, T+t<T_{\varphi}} \|t^{1-\varepsilon'-\delta}A^{1-\varepsilon'-\delta}e^{-tA}u(T)\|$$

is not uniformly convergent to O _on_ $T \in [O,T_{\varphi})$ _as_ $\tilde{\delta} \to O$ _for every_ ε', $O < \varepsilon' < 1-\delta$.

Proof: We set $w(t) = A^{-\delta}u(t)$. We want to apply theorem II.3.6. For this purpose the mapping $A^{-\delta}\tilde{M}(.)$ is continued to a mapping from $D(A^{1-\rho_1'})$ into B satisfying the Lipschitz conditions indicated in theorem II.3.8. This continuation is also denoted by $A^{-\delta}\tilde{M}(.)$. Then the assumptions of theorem II.3.6 are fulfilled with respect to the integral equation

$$w(t) = e^{-tA}A^{-\delta}\varphi - \int_{O}^{t} e^{-(t-s)A}A^{-\delta}\tilde{M}(w(s))\,ds;$$

to see this we have to replace M(u) by the nonlinearity $A^{-\delta}\tilde{M}(u)$, ρ_1 by ρ_1', ρ_o is the number 1 and according to our assumption we have

$$1 - \rho_1' - \frac{\rho_1'}{\rho_o} = 1 - 2\rho_1' \leqq \delta < 1-\rho_1'.$$

Thus theorem II.3.8 follows up to (II.3.11). The characterisation of the case $T_{\varphi} < +\infty$ in theorem II.3.6 immediately furnishes

$$\lim_{t \uparrow T_\varphi} \|A^{1-\rho_1'} w(t)\| = \infty.$$

As for the last statement of theorem II.3.8 it follows from the construction of $\mathfrak{M}_{\tilde{T}}$ in the proof of theorem II.3.6 that for $T_\varphi <$ $+\infty$: $\lim_{t \uparrow T_\varphi} \|A^\delta w(t)\| = +\infty$ or $\sup_{0 \le t \le \tilde{\delta}, T+t<T_\varphi} \|t^{1-\rho_1'-\delta} A^{1-\rho_1'} e^{-tA} w(T)\|$ is not uniformly convergent to 0 as $\tilde{\delta} \to 0$.

Because of $\|t^{1-\rho_1'-\delta} A^{1-\rho_1'} e^{-tA} w(T)\| \le \|t^{-\rho_1'+\varepsilon'} A^{-\rho_1'+\varepsilon'} e^{-(t/2)A} t^{1-\varepsilon'-\delta} A^{1-\varepsilon'} e^{-(t/2)A} w(T)\| \le c \cdot \|t^{1-\varepsilon'-\delta} A^{1-\varepsilon'} e^{-(t/s)A} w(T)\|$ our theorem is proved. □

We now turn to the question whether if our results II.3.3, II.3.4, II.3.5, II.3.6,and II.3.7 remain valid if an inhomogeneity $f(t)$ is added to $M(u)$. To make it more precise, we want to give suitable regularity criteria concerning f in order that the theorems mentioned above remain true.

Theorem II.3.9: 1. Let ρ_1 be the quantity from theorems II.3.3 and II.3.4. Let $f \in L^q((0,T),B)$ for some $q > \frac{1}{\rho_1}$ and for all $T > 0$. Then the statements of theorems II.3.3 and II.3.4 remain true for the integral equation

$$\text{(II.3.12)} \quad u(t) = e^{-tA}\varphi + \int_0^t e^{-(t-s)A}(-M(u(s))+f(s))\, ds.$$

2. Let $f \in C^0([0,T], D(A^{\kappa+\rho_2-1}))$ for all $T > 0$. Then the statement of theorem II.3.5 remains true for the integral equation (II.3.12).

3. Let $f \in L^q((0,T),B)$ for all $T > 0$ and for some $q > \frac{1}{\rho_1}$ where ρ_1 is the quantity from theorem II.3.6. Then the statement of theorem II.3.6 remains true for the integral equation (II.3.12). If $f \in C^0([0,T], D(A^{\kappa+\rho_2-1}))$ then the statement of theorem II.3.7 remains true for the integral equation (II.3.12).

4. Let δ be the quantity from theorem II.3.8, let $f \in L^{\frac{1}{\delta}}((0,T),B)$, $T > 0$. Then the statement of theorem II.3.8 remains true for the integral equation (II.3.12).

5. If $f \in L^{\infty}((0,T),B)$ then we have for the solution of the equation (II.3.12) in the cases 1. and 3.:

$$u(t) \in \bigcap_{0<\varepsilon<1} D(A^{1-\varepsilon}(t)), \quad 0 < t < T_{\varphi},$$

$$A^{1-\varepsilon}u \in C^{0}((0,T_{\varphi}),B), \quad 0 < \varepsilon < 1.$$

Proof: If $f \in L^{q}((0,T),B)$ for some $q > \frac{1}{\rho_1} = 1/(1-\rho_1)/(\frac{1}{1-\rho_1}-1)$ then

$$\text{(II.3.13)} \quad \int_0^t e^{-(t-s)A} f(s)\, ds \in D(A^{1-\rho_1}), \quad 0 \leq t \leq T,$$

and

$$A^{1-\rho_1} \int_0^{\cdot} e^{-(\cdot-s)A} f(s)\, ds \in L^{\infty}((0,T),B).$$

As for the continuity of the integral term we have for $t > t'$

$$\|A^{1-\rho_1} \int_0^t e^{-(t-s)A} f(s)\, ds - \int_0^{t'} e^{-(t'-s)A} f(s)\, ds\|$$

$$\leq \|\int_{t'}^t A^{1-\rho_1} e^{-(t-s)A} f(s)\, ds\| + \|\int_0^{t'} A^{1-\rho_1} (e^{-(t-s)A} - e^{-(t'-s)A}) f(s)\, ds\|.$$

Using Hölder's inequality with $p_1 = \frac{q}{q-1}$, $p_2 = q$ for the first term we see that it converges to 0 for $t \to t'$. As for the second term we have according to (II.3.3)

$$\|A^{1-\rho_1}(e^{-(t-s)A} - e^{(t'-s)A})\| \leq c|t-t'|^{\tilde{\delta}}/|t-s|^{1-\rho_1+\tilde{\delta}}$$

for any $\tilde{\delta}$, $1-\rho_1 \geq \tilde{\delta} \geq 0$. For $\tilde{\delta}$ sufficiently small we again apply Hölder's inequality with the same exponents as before. This proves the first part.

As for the second part we simply remark that

$$A^{\kappa-(1-\rho_2)+1-\hat{\eta}} \int_0^t e^{-(t-s)A} f(s)\, ds \in C^0([0,T],B), \quad \hat{\eta} > 0,$$

if $f \in C^0([0,T],A^{\kappa-(1-\rho_2)})$. The proof is like that of the first part of theorem II.3.5 and may be left to the reader.

As for the third part we refer to the proof of part 1.,2. The fourth part simply follows from the fact that for $0 < \varepsilon' < 1$

$$\left\| \int_0^t A^{1-\varepsilon'-\delta} e^{-(t-s)A} f(s)\, ds \right\|$$

$$\leq \int_0^t \frac{c}{(t-s)^{1-\varepsilon'-\delta}} \|f(s)\|\, ds$$

$$\leq \left(\int_0^t \frac{c}{(t-s)^{(1-\varepsilon'-\delta)/(1-\delta)}}\right)^{1-\delta} \left(\int_0^t \|f(s)\|^{1/\delta}\, ds\right)^{\delta}$$

and that

$$A^{1-\varepsilon'-\delta} \int_0^{\cdot} e^{-(\cdot-s)A} f(s)\, ds \in C^0([0,T],B).$$

The latter is readily seen by using (II.3.3). For the fifth part we refer to the last lines of the proof of theorem II.3.7. □

We now study the continuity properties of a solution of u of $u' + Au + M(u) = f$ if $t \to 0$ provided the initial value has a comparatively high degree of regularity. This will be used later on to study the regularity up to $t = 0$ of strong solutions of the Navier-Stokes equations without assuming higher order compatibility conditions in $t = 0$ (which may be difficult to verify). The next theorem is different in character from the preceding ones: The initial value is in the domain of definition of the same power of A as the solution for $t > 0$. It is for this that we have separated the following result from the preceding theory.

Theorem II.3.10: *Let* M *fulfil the assumptions of theorem* II.3.1. *Let* $\varphi \in D(A^{\kappa+\rho_2})$. *Then the solution of* $u' + Au + M(u) = 0$, $u(0) = \varphi$, *in theorem* II.3.1 *has the following properties*:

$$u(t) \in D(A^{\kappa+\rho_2}),$$

$$A^{\kappa+\rho_2}u(.) \in C^{o}([0,T_{\varphi}),B),$$

$$u'(.) \in D(A^{\kappa+\rho_2-1}),$$

$$A^{\kappa+\rho_2-1}u'(.) \in C^{o}([0,T_{\varphi}),B).$$

If we choose some $f \in C^{o}([0,T],B)$, $T > 0$, *such that*

$$\int_0^{\cdot} e^{-(.-s)A}f(s)\,ds \in D(A^{\kappa+\rho_2}),\ t \geq 0,$$

$$A^{\kappa+\rho_2}\int_0^{\cdot} e^{-(.-s)A}f(s)\,ds \in C^{o}([0,T],B),\ T > 0,$$

then the statement above remains true for the solution of $u' + Au + M(u) = f$, $u(0) = \varphi$.

Proof: We already know that $u(t) \in D(A^{\kappa})$, $0 \leq t \leq \tilde{T} < T_{\varphi}$,

$$A^{\kappa}u(.) \in \bigcap_{0<\tilde{T}<T_{\varphi}} C^{o}([0,\tilde{T}],B).$$

M being from $A(D(A^{\kappa}),D(A^{\kappa-(1-\rho_2)}))$ we have

$$\|A^{\kappa-(1-\rho_2)}(M(u(t))-M(u(t')))\|$$

$$\leq g(\|A^{\kappa}u(t)\| + \|A^{\kappa}u(t')\|)\|A^{\kappa}(u(t)-u(t'))\|$$

with some continuous function g. Moreover

$$A^{\kappa}\int_0^t e^{-(t-s)A}M(u(s))\,ds$$

$$=\int_0^t A^{\kappa}e^{-(t-s)A}A^{-(\kappa-(1-\rho_2))}A^{\kappa-(1-\rho_2)}M(u(s))\,ds$$

and

$$\|A^{\kappa}e^{-(t'-s)A}A^{-(\kappa-(1-\rho_2))}-A^{\kappa}e^{-(t-s)A}A^{-(\kappa-(1-\rho_2))}\|$$

$$\leq c\,\frac{|t'-t|^{\tilde{\eta}}}{|t-s|^{1-\rho_1+\tilde{\eta}}}\quad\text{for some } \tilde{\eta}>0$$

(cf. (II.3.3)). From this it follows that

$$A^{\kappa}\int_0^{\cdot} e^{-(\cdot-s)A}M(u(s))\,ds \in C^{\tilde{\eta}}([0,\tilde{T}],B),\ \tilde{T}<T_{\varphi},$$

for some $\tilde{\eta}>0$. From this we see that the integral term is in $D(A^{\kappa+\rho_2})$ and

$$A^{\kappa+\rho_2}\int_0^{\cdot} e^{-(\cdot-s)A}M(u(s))\,ds \in C^{0}([0,\tilde{T}],B),\ \tilde{T}<T_{\varphi}.$$

Our assumption on φ completes the proof.

§ 4. Comments to Chapter II

To § 1: Theorem II.1.1 is stated in [K2, pp. 56,57] without a proof. A proof in the case that A generates a continuous semi-group e^{-tA} is given in [W3, I]; there also the time dependent case

$$u'+A(t)u+M(t,u)=0$$

is treated and moreover the consequences for nonlinear wave equations

$$u''(t)+A(t)u+M(t,u,u')=0$$

are treated in detail with some applications to problems of non-linear elasticity. These are also considered in [CW].

To § 2: The basic idea of the nonlinear interpolation theory given here goes back to Heinz [H2]; Heinz dealt with the case B = Hilbert space and $A(t)$ = selfadjoint only. The general case was treated in [W1, W2]. In the selfadjoint case we are allowed to set $\eta = 0$ since the pure imaginary powers $A^{i\tau}(t)$ are uniformly bounded. In fact if $D(A^{\gamma}(t))$ is the complex interpolation space $[B, D(A^{l}(t))]_{\gamma/l}$, $l > \gamma$, $l \in \mathbb{N}$, then we also have $\eta = 0$; this follows from a recent paper of J. Bergh [Be]; $D(A^{\gamma}(t)) = [B, D(A^{l}(t))]_{\gamma/l}$ is by now known in many important cases, e.g.: B is an L^{p}-space, $A(t)$ a second order elliptic operator with suitable boundary conditions ([Tr1, 4.9]) or the Stokes operator ([Gi1]).

To § 3: The theory developed here remains still valid for equations of the form

$$u' + A(t)u + M(u) = 0,$$
$$u(0) = \varphi.$$

Here each of the $-A(t)$, $0 \leq t \leq T$, is the generator of an analytic semigroup (see [W2]); moreover we assume that at least each $A(t)$ has constant domain of definition $D(A(t))$. It may be also necessary to assume that a certain $A^{k_0}(t)$, $k_0 \in \mathbb{N}$, has constant domain of definition. Moreover we assume that $A(t)x$ or $A^{k_0}(t)x$, $0 \leq t \leq T$, $x \in D(A(0))$ or $x \in D(A^{k_0}(0))$, is twice continuously differentiable (this assumption may be weakened). On the basis of suitable estimates for $\|A^{\tilde{\gamma}+\gamma}(t)U(t,s)A^{-\gamma}(s)\|$ and $\|\frac{d}{dt}(A^{\gamma}(t)x)\|$ for certain (possibly fractional) $\tilde{\gamma}, \gamma$ practically all results of § 3 can be carried over to the time dependent case; $U(t,s)$ denotes the evolution operator in the sense of theorem I.2.1. As announced in the proof of theorem II.3.4 we want to give a description of Kielhöfer's Lemma which is useful in various situations: Let $\varphi \in C^{0}([t_0, T_0))$ for some t_0, T_0, $0 \leq t_0 < T_0$, let $\varphi(t) \geq 0$ on $[t_0, T_0)$ and let φ satisfy the inequality

$$\varphi(t) \leq \tilde{c} + \tilde{c} \int_{t_o}^{t} (t-s)^{-\frac{1}{2}-\eta} \varphi(s)^{2(1-\eta)} ds, \quad t_o \leq t < T_o,$$

for some $\tilde{c} \geq 0$ and some $\eta \in [0, \frac{1}{2})$. Let $\varphi \in L^2((t_o, T_o))$. Then $\varphi(t) \leq L(\tilde{c}, \eta, \varphi(t_o), \|\varphi\|_{L^2((t_o,T_o))})$. A proof can be found in [Ki1, p. 218].

III. Local Solvability of the Equations of Navier-Stokes

First in this chapter we want to discuss properties of the linear part of the Navier-Stokes equations. Then we draw some consequences in order to solve these equations locally in time with "bad" initial values. Our main tool are the results of II.

§ 1. Solonnikov's Results for the Instationary Stokes Equation

To avoid technical difficulties we assume from now throughout this book that Ω is a bounded open and pathwise connected set of $\mathbb{R}^n$ of class C^∞. In what follows we concentrate on n=2 or n=3. As Solonnikov ([Sol 2]) has proved the following theorem holds:

<u>Theorem III.1.1:</u> <u>Let</u> $p > 1$, $p \neq 3/2$, $T > 0$, $f \in L^p((0,T),(L^p(\Omega))^n)$, $\varphi \in (\tilde{W}^{2-2/p,p}(\Omega))^n \cap H_p(\Omega)$. [1] <u>Then there is one and only one</u>

$$u \in L^p((0,T),(H^{2,p}(\Omega))^n) \cap L^p((0,T),(\overset{\circ}{H}{}^{1,p}(\Omega))^n)$$

<u>with</u>

$$u' \in L^p((0,T),(L^p(\Omega))^n)$$

$$u \in C^0([0,T],(\tilde{W}^{2-2/p,p}(\Omega))^n)$$

<u>and one</u> $\pi \in L^p((0,T),L^p(\Omega))$ <u>with</u>

$$\nabla\pi \in L^p((0,T),(L^p(\Omega))^n)$$

<u>such that</u>

---

[1] For $H_p(\Omega)$ cf. the auxiliary propositions in 0.2. In what follows ν is a positive constant (the viscosity).

$$(\mathrm{III.1.1})\quad \begin{cases} u' - \nu\Delta u + \nabla\pi = f, \\ \nabla\cdot u = 0, \\ u(0) = \varphi. \end{cases}$$

π is uniquely determined up to a function of t being in $L^p((0,T))$. Moreover the following estimate holds

$$(\mathrm{III.1.2})\quad \int_0^T \|u'(t)\|^p_{(L^p(\Omega))^n}\,dt + \int_0^T \|u(t)\|^p_{(H^{2,p}(\Omega))^n}\,dt +$$

$$+ \sup_{0\le t\le T} \|u(t)\|^p_{(W^{2-2/p,p}(\Omega))^n} + \int_0^T \|\nabla\pi(t)\|^p_{(L^p(\Omega))^n}\,dt$$

$$\le c(\nu,T)\left(\int_0^T \|f(t)\|^n_{(L^p(\Omega))^n}\,dt + \|\varphi\|_{(W^{2-2/p,p}(\Omega))^n}\right).$$ [1]

We cannot prove this important theorem here but we will make some remarks on it. First we concentrate on the mathematical aspects of (III.1.1), (III.1.2). The estimate (III.1.2) is similar to the well known estimate for parabolic equations which can be found in the book of Ladyženskaja, Ural'ceva and Solonnikov [LUS, IV] although the system (III.1.1) is not parabolic because of the presence of $\nabla\pi$ and the equation $\nabla\cdot u = 0$; u and π are the unknown quantities in (III.1.1) and we thus have n+1 equations for the n components of u and the scalar function π.

$c(\nu,T)$ still depends on T and it has the property that $c(\nu,T') \le c(\nu,T'')$ if $T' \le T''$. For us it is important to replace $c(\nu,T)$ by a quantity $c(\nu)$ which does not depend on T. This is done in the following way: Let

$$\xi_0(t) = \begin{cases} 1, & 0\le t\le 1, \\ -3t+4, & 1\le t\le 1+\frac{1}{3} \\ 0, & t\ge 1+\frac{1}{3}, \end{cases}$$

---

[1] For the property: $\pi \in L^p((0,T),L^p(\Omega))$ cf. the auxiliary propositions in 0.2 on p. XVIII.

$$\xi_\nu(t) = \begin{cases} 0, & t \le \nu-\frac{1}{3} \\ 3t+(1-3\nu), & \nu-\frac{1}{3} \le t \le \nu, \\ 1, & \nu \le t \le \nu+1, \\ -3t+1+3(\nu+1), & \nu+1 \le t \le \nu+1+\frac{1}{3}, \\ 0, & t \ge \nu+1+\frac{1}{3}. \end{cases}$$

Then $\sum_{\nu=0}^{\infty} \xi_\nu(t) \le 2$, and an arbitrary $t \ge 0$ is contained at most in the support of two ξ_ν, $\nu \in \mathbb{N} \cup \{0\}$. Let $T \in (\nu_0, \nu_0+1)$ for some $\nu_0 \in \mathbb{N} \cup \{0\}$. For every μ, $1 \le \mu$ we have

$$(\xi_\mu u)' - \nu\Delta(\xi_\mu u) + \nabla\xi_\mu \pi = \xi_\mu f + \xi_\mu' u$$

and therefore by (III.1.2):

$$\text{(III.1.3)} \quad \int_\mu^{\mu+1} \|u'(t)\|^p_{(L^p(\Omega))^n} + \int_\mu^{\mu+1} \|u(t)\|^p_{(H^{2,p}(\Omega))^n} dt + $$
$$+ \sup_{\mu \le t \le \mu+1} \|u(t)\|^p_{(W^{2-2/p,p}(\Omega))^n} + \int_\mu^{\mu+1} \|\nabla\pi(t)\|^n_{(L^p(\Omega))^n} dt$$
$$\le c(\nu, 1+\tfrac{2}{3})\Big(\int_{\mu-\frac{1}{3}}^{\mu+1+\frac{1}{3}} \|f(t)\|^p_{(L^p(\Omega))^n} dt + $$
$$+ \int_{\mu-\frac{1}{3}}^{\mu+1+\frac{1}{3}} \|u(t)\|^p_{(L^p(\Omega))^n} dt\Big).$$

The same argument shows that

$$\int_{\nu_0}^{T} \|u'(t)\|^p_{(L^p(\Omega))^n} dt + \int_{\nu_0}^{T} \|u(t)\|^p_{(H^{2,p}(\Omega))^n} dt + $$
$$+ \sup_{\nu_0 \le t \le T} \|u(t)\|^p_{(W^{2-2/p,p}(\Omega))^n} + \int_{\nu_0}^{T} \|\nabla\pi(t)\|^p_{(L^p(\Omega))^n} dt$$
$$\le c(\nu,1)\Big(\int_{\nu_0-\frac{1}{3}}^{T} \|f(t)\|^p_{(L^p(\Omega))^n} dt + \int_{\nu_0-\frac{1}{3}}^{T} \|u(t)\|^p_{(L^p(\Omega))^n} dt\Big),$$

and (if $\nu_o \geq 1$)

$$\int_0^1 \|u'(t)\|^p_{(L^p(\Omega))^n}\, dt + \int_0^1 \|u(t)\|^p_{(H^{2,p}(\Omega))^n}\, dt +$$

$$+ \sup_{0\leq t\leq 1} \|u(t)\|^p_{(W^{2-2/p,p}(\Omega))^n} + \int_0^1 \|\nabla\pi(t)\|^p_{(L^p(\Omega))^n}\, dt$$

$$\leq c(\nu,1+\tfrac{1}{3})\,(\int_0^{1+\frac{1}{3}} \|f(t)\|^p_{(L^p(\Omega))^n}\, dt + \|\varphi\|_{(W^{2-2/p,p}(\Omega))^n} +$$

$$+ \int_0^{1+\frac{1}{3}} \|u(t)\|^p_{(L^p(\Omega))^n}\, dt).$$

Adding all these inequalities we get

$$\text{(III.1.4)}\quad \int_0^T \|u'(t)\|^p_{(L^p(\Omega))^n}\, dt + \int_0^T \|u(t)\|^p_{(H^{2,p}(\Omega))^n}\, dt +$$

$$+ \sup_{0\leq t\leq T} \|u(t)\|^p_{(W^{2-2/p,p}(\Omega))^n} + \int_0^T \|\nabla\pi(t)\|^p_{(L^p(\Omega))^n}\, dt$$

$$\leq c(\nu)\,(\int_0^T \|f(t)\|^p_{(L^p(\Omega))^n}\, dt + \|\varphi\|_{(W^{2-2/p,p}(\Omega))^n} +$$

$$+ \int_0^T \|u(t)\|^p_{(L^p(\Omega))^n}\, dt)$$

with some constant $c(\nu)$ being independent from T. Proof and result are the same if T is one of the points ν_o or ν_o+1. The expression on the right side of (III.1.3) or (III.1.4) is denoted by

$$|||(u,\pi)|||^p_{p,(\mu,\mu+1)},\quad |||(u,\pi)|||^p_{p,(0,T)}$$

respectively. (III.1.4) can still be improved: We have

$$\|u(t)\|_{(L^p(\Omega))^n} \leq \varepsilon\|u(t)\|_{(H^{2,p}(\Omega))^n} + c(\varepsilon)\|u(t)\|_{(L^1(\Omega))^n}$$

for all $\varepsilon > 0$ (cf. our lemma 0.2.2). Thus we get

(III.1.5) $$|||(u,\pi)|||^{p}_{p,(0,T)} \leq$$

$$\leq c(\nu)\left(\int_0^T \|f(t)\|^p_{(L^p(\Omega))^n}\, dt + \int_0^T \|u(t)\|^p_{(L^1(\Omega))^n}\, dt + \right.$$

$$\left. + \|\varphi\|^p_{(W^{2-2/p,p}(\Omega))^n}\right).$$

Now let $p \geq 2$. Multiplying the equation $u' - \nu\Delta u + \nabla\pi = f$ scalarly by u we get

$$\int_0^{\tilde{T}} (u'(t),u(t))\, dt + \nu \int_0^{\tilde{T}} \|\nabla u(t)\|^2_{(L^2(\Omega))^{n^2}}\, dt =$$

$$= \int_0^{\tilde{T}} (f(t),u(t))\, dt, \quad 0 \leq \tilde{T} \leq T,$$

since by 0.2. we have $(\nabla\pi(t),u(t)) = 0$. We get

(III.1.6) $$u' \in L^2((0,T),(L^2(\Omega))^n) \subset L^2((0,T),(H^{-1,2}(\Omega))^n),$$

$$u \in L^2((0,T),(\overset{\circ}{H}{}^{1,2}(\Omega))^n).$$

(0.2.6) shows that

$$\|u(\tilde{T})\|^2_{(L^2(\Omega))^n} - \|\varphi\|^2_{(L^2(\Omega))^n} + 2\nu\int_0^{\tilde{T}} \|\nabla u(t)\|^2_{(L^2(\Omega))^{n^2}}\, dt$$

$$= 2\,\mathrm{Re} \int_0^{\tilde{T}} (f(t),u(t))\, dt$$

$$\leq \varepsilon \int_0^{\tilde{T}} \|\nabla u(t)\|^2_{(L^2(\Omega))^n}\, dt + c(\varepsilon) \int_0^{\tilde{T}} \|f(t)\|_{(L^2(\Omega))^n}\, dt,$$

$\varepsilon > 0$. This gives

$$(III.1.7)\quad \|u(\tilde{T})\|^2_{(L^2(\Omega))^n} + (2\nu-\varepsilon)\int_0^{\tilde{T}} \|\nabla u(t)\|^2_{(L^2(\Omega))^n}$$

$$\leq \|\varphi\|^2_{(L^2(\Omega))^n} + c(\varepsilon)\int_0^{\tilde{T}} \|f(t)\|^2_{(L^2(\Omega))^n}\, dt.$$

Inserting (III.1.7) into (III.1.5) we arrive at

$$|||(u,\pi)|||^p_{p,(0,T)} \leq c(\nu)\Big(\int_0^T \|f(t)\|^p_{(L^p(\Omega))^n}\, dt +$$

$$+ \Big(\int_0^T \|f(t)\|^2_{(L^2(\Omega))^n}\, dt\Big)^{p/2} + \|\varphi\|^p_{(W^{2-2/p,p}(\Omega))^n}\Big).$$

Observe that all constants still depend on Ω.

After having dealt with the linear instationary problem we make a few remarks on the stationary problem

$$(III.1.8)\quad \begin{cases} -\nu\Delta u + \nabla\pi = f, \\ \nabla\cdot u = 0, \\ u|\partial\Omega = 0. \end{cases}$$

It is well known since several years [KS, P. 307] that this problem is elliptic in the sense of [ADN II, Theorem 10.5]. As Témam has proved [Tem, p. 33] it follows from [ADN II, Theorem 10.5] that the following a-priori estimate holds: Let $p > 1$, $f \in (L^p(\Omega))^n$. Let $u \in (H^{2,p}(\Omega))^n \cap (\overset{\circ}{H}{}^{1,p}(\Omega))^n$, $\pi \in H^{1,p}(\Omega)$ fulfill (III.1.8). Then

$$(III.1.9)\quad \|u\|_{(H^{2,p}(\Omega))^n} + \|\nabla\pi\|_{(L^p(\Omega))^n}$$

$$\leq c(\|f\|_{(L^p(\Omega))^n} + \|u\|_{(L^p(\Omega))^n} + \|\pi\|_{L^p(\Omega)}).$$

If $f \in (H^{l,p}(\Omega))^n$ for some $l \in \mathbb{N}$ then $u \in (H^{2+l,p}(\Omega))^n$, $\nabla\pi \in (H^{l,p}(\Omega))^n$ and the right side of (III.1.9) may be replaced by $\|u\|_{(H^{2+l,p}(\Omega))^n} + \|\nabla\pi\|_{(H^{l,p}(\Omega))^n}$, the left side by $\|f\|_{(H^{l,p}(\Omega))^n} +$

$+ \|u\|_{(L^p(\Omega))^n} + \|\pi\|_{L^p(\Omega)}$) (cf. [ADN II, theorem 10.5]).

As it is proved in [L, 3.5], [Sol 2], in fact the stronger estimate

$$\text{(III.1.10)} \quad \|u\|_{(H^{2+l,p}(\Omega))^n} + \|\nabla\pi\|_{(H^{l,p}(\Omega))^n} \le c\|f\|_{(H^{l,p}(\Omega))^n}, \quad l \in \mathbb{N},$$

holds. Thus a solution (u,π) of (III.1.8) being in the class above is determined uniquely in the sense that u is determined uniquely and π is determined uniquely up to a constant. It is also proved in [L, 3.5] that a solution of (III.1.8) with $u \in (H^{2,p}(\Omega))^n \cap (\overset{o}{H}{}^{1,p}(\Omega))^n$, $\pi \in H^{1,p}(\Omega)$ also exists. This corresponds to theorem III.1.1, and moreover it is a complete analogon to the well known estimates for the graph-norm of the Laplacian in L^p-spaces.

Now let $P = P_p$ be the "projection" of $(L^p(\Omega))^n$ on its divergence free part as it has been introduced in 0.2. We set

$$Au = A_p u = -\nu P_p \Delta u, \quad u \in (H^{2,p}(\Omega))^n \cap (\overset{o}{H}{}^{1,p}(\Omega))^n,$$
$$\nabla\cdot u = 0.$$

Then it already follows from (III.1.10) that A_p is closed in $H_p(\Omega)$ with dense domain of definition $D(A) = \{u | u \in H^{2,p}(\Omega))^n \cap (\overset{o}{H}{}^{1,p}(\Omega))^n, \nabla\cdot u = 0\}$. At this point we use the fact that

$$\overset{o}{H}{}^{1,p}_{*}(\Omega) = \{u | u \in (\overset{o}{H}{}^{1,p}(\Omega))^n, \nabla\cdot u = 0\}$$
$$= \overline{\{u | u \in (C_o^\infty(\Omega))^n, \nabla\cdot u = 0\}}^{\|\cdot\|_{(H^{1,p}(\Omega))^n}},$$
$$\subset H_p(\Omega);$$

this will be proved in the next paragraph (independently, from our considerations here). In what follows we will show that $-A_p$, $p > 1$, generates an analytic semigroup. Moreover, if $p = 2$, A_2 is a positive selfadjoint operator. We will start with the second statement:

<u>Theorem III.1.2:</u> $A = A_2$ <u>is a positive selfadjoint operator in the Hilbert space</u> $H_2(\Omega)$.

<u>Proof:</u> We have

$$(Au,w) = (-\nu\Delta u,w) = (u,Aw),$$
$$(Au,u) = \nu\|\nabla u\|^2_{(L^2(\Omega))^n}, \quad w,u \in D(A).$$

$(\cdot,\cdot)$ is the scalar product in $(L^2(\Omega))^n$ as well as in $H_2(\Omega)$. Thus A is positive symmetric. As we have remarked before, for any $f \in (L^2(\Omega))^n$ there is a unique $u \in D(A)$ and a unique (up to a constant) π such that

$$-\nu\Delta u + \nabla\pi = f,$$
$$\nabla\cdot u = 0.$$

On applying $P = P_2$ to the first equality we see that $Au = Pf$. Thus $R(A) = H_2(\Omega)$ and A is selfadjoint.

Next we want to show that $-A$ generates an analytic semigroup in $H_p(\Omega)$.

<u>Theorem III.1.3:</u> <u>Let</u> $p > 1$, $p \neq \frac{3}{2}$. <u>There is a</u> $\Lambda_o < 0$ <u>such that for</u> $\lambda \in \mathbb{C}$, Re $\lambda \geq \Lambda_o$ <u>there is a unique</u> $u \in D(A_p)$ <u>with</u>

$$-\nu P_p\Delta u + \lambda u = Pf.$$

<u>Moreover the operator</u> $(-\nu P_p\Delta+\lambda)^{-1} = (A_p+\lambda)^{-1}$ <u>satisfies the estimate</u>

$$\|(A_p+\lambda)^{-1}\| \leq \frac{c}{|\lambda|+1}, \quad \|A_p(A_p+\lambda)^{-1}\| \leq c,$$

$$\frac{1}{2}\pi+\varepsilon \geq \arg(\lambda-\Lambda_o) \geq \frac{3}{2}\pi-\varepsilon,$$

<u>for some</u> $\varepsilon > 0$ <u>with</u> $0 < \varepsilon < \frac{\pi}{2}$.

Proof: The basic idea of the proof is due to Sobolevskii [Sb, pp. 720-721]. Let $\lambda \in \mathbb{C}$, $u_o \in D(A_p)$, $\pi_o \in H^{1,p}(\Omega)$, $f_o \in L^p(\Omega)$

$$-\nu\Delta u_o + \lambda u_o + \nabla\pi_o = f_o, \quad \nabla\cdot u_o = 0.$$

We set

$$u(t,x) = e^{\lambda t}u_o(x),$$
$$\pi(t,x) = e^{\lambda t}\pi_o(x),$$
$$f(t,x) = e^{\lambda t}f_o(x).$$

Then we get

$$u' - \nu\Delta u + \nabla\pi = e^{\lambda t}f_o,$$

$$|||(u,\pi)|||^p_{p,(0,T)} \le c(\|u_o\|^p_{(H^{2,p}(\Omega))^n} + \int_0^T \|e^{\lambda t}f_o\|^p_{(L^p(\Omega))^n}\,dt + \int_0^T \|u(t)\|^p_{(L^p(\Omega))^n}\,dt).$$

Here we have applied (III.1.4). Since $u'(t,x) = \lambda e^{\lambda t}u_o(x)$ it follows that

$$|\lambda|^p \int_0^T e^{p\,\mathrm{Re}\,\lambda t}\|u_o\|^p_{(L^p(\Omega))^n}\,dt + \int_0^T e^{p\,\mathrm{Re}\,\lambda t}\|u_o\|^p_{(H^{2,p}(\Omega))^n}\,dt$$

$$\le c(\|u_o\|^p_{(H^{2,p}(\Omega))^n} + \int_0^T e^{p\,\mathrm{Re}\,\lambda t}\|f_o\|^p_{(L^p(\Omega))^n}\,dt + \int_0^T e^{p\,\mathrm{Re}\,\lambda t}\|u_o\|^p_{(L^p(\Omega))^n}\,dt),$$

$$\frac{e^{p\,\mathrm{Re}\,\lambda T}-1}{p\,\mathrm{Re}\,\lambda}(|\lambda|^p\|u_o\|^p_{(L^p(\Omega))^n} + \|u_o\|^p_{(H^{2,p}(\Omega))^n})$$

$$\le c(\|u_o\|^p_{(H^{2,p}(\Omega))^n} + \frac{e^{p\,\mathrm{Re}\,\lambda T}-1}{p\,\mathrm{Re}\,\lambda}(\|f_o\|^p_{(L^p(\Omega))^n} + \|u_o\|^p_{(L^p(\Omega))^n}).$$

Now let $\text{Re}\,\lambda > 0$. If $|\lambda|^p \geq c+1$, where c is the constant on the right side of the last inequality, we get with $T \geq c+1$, $(e^{p\,\text{Re}\,\lambda\,T} - 1)/p\,\text{Re}\,\lambda \geq T \geq c+1$, the inequalities

$$\frac{e^{p\,\text{Re}\,\lambda\,T}-1}{p\,\text{Re}\,\lambda}(|\lambda|^p - c)\|u_o\|^p_{(L^p(\Omega))^n} + \left(\frac{e^{p\,\text{Re}\,\lambda\,T}-1}{p\,\text{Re}\,\lambda} - c\right)\|u_o\|^p_{(H^{2,p}(\Omega))^n}$$

$$\leq c\cdot\frac{e^{p\,\text{Re}\,\lambda\,T}-1}{p\,\text{Re}\,\lambda}\|f_o\|^p_{(L^p(\Omega))^n},$$

$$(|\lambda|^p - c)\|u_o\|^p_{(L^p(\Omega))^n} + \left(1 - \frac{c}{\frac{e^{p\,\text{Re}\,\lambda\,T}-1}{p\,\text{Re}\,\lambda}}\right)\|u_o\|^p_{(H^{2,p}(\Omega))^n}$$

$$\leq c\|f_o\|^p_{(L^p(\Omega))^n},$$

$$\text{(III.1.11)}\quad (|\lambda|^p - c)\|u_o\|^p_{(L^p(\Omega))^n} + \frac{1}{c+1}\|u_o\|_{(H^{2,p}(\Omega))^n} \leq$$

$$\leq c\|f_o\|^p_{(L^p(\Omega))^n}.$$

Observe that the constants in this estimate do not depend on T. We already know that for $\text{Re}\,\lambda \geq \Lambda_o$ with some $\Lambda_o < 0$ the resolvent $(\lambda+A_2)^{-1}$ exists in $H_2(\Omega)$: Using the spectral resolution $E(\lambda)$ of A it follows that for some $\varepsilon > 0$

$$(\lambda+A_2)^{-1}x = \int_\varepsilon^\infty \frac{1}{\lambda+\mu}dE(\mu)x,$$

$$A(\lambda+A_2)^{-1}x = \int_\varepsilon^\infty \frac{\mu}{\lambda+\mu}dE(\mu)x;$$

thus we have

$$\text{(III.1.12)}\quad \|(\lambda+A_2)^{-1}\| \leq \frac{M'}{|\lambda|+1},\quad \text{Re}\,\lambda \geq \Lambda_o,$$

$$\text{(III.1.13)}\quad \|A_2(\lambda+A_2)^{-1}\| \leq M'',\quad \text{Re}\,\lambda \geq \Lambda_o.$$

Now let $p \geq 2$. We want to solve $(\lambda+A_p)u = f$ for any $f \in H_p(\Omega)$, and λ with $\mathrm{Re}\ \lambda \geq 0$. We already know that there is a unique $u \in D(A_2)$ with $(\lambda+A_2)u = f$. By Sobolev (cf. our lemma 0.2.1) we have

$$u \in (C^o(\bar{\Omega}))^n \cap H_p(\Omega),$$

$$\|u\|_{(C^o(\bar{\Omega}))^n} \leq c\|A_2 u\|.$$

Thus

$$A_2 u = -\lambda u + f \in H_p(\Omega).$$

As mentioned before, from [L, 3.5] it follows that $u \in D(A_p)$. If $p < 2$, $p \neq \frac{3}{2}$, we approximate $f \in H_p(\Omega)$ in $H_p(\Omega)$ by $f_\nu \in C_o^\infty(\Omega)$ with $\nabla \cdot f_\nu = 0$, $\nu \in \mathbb{N}$. There is already a solution of $(\lambda+A_p)u_\nu = f_\nu$ if $\mathrm{Re}\ \lambda > 0$. If $|\lambda| \geq {}^p\sqrt{c+1}$ the inequality (III.1.11) shows that the u_ν converge to the desired solution u in $(H^{2,p}(\Omega))^n$.

Thus we get with (III.1.11) that the resolvent $(\lambda+A_p)^{-1}$ exists $\mathrm{Re}\ \lambda > 0$, $|\lambda| \geq {}^p\sqrt{c+1}$, $p > 1$, $p \neq \frac{3}{2}$. Moreover we have the estimate

$$\|(\lambda+A_p)^{-1}\| \leq \frac{c}{|\lambda|-c'}.$$

Observe that c' only depends on c, and c does not depend on λ. Using the resolvent series (cf. I.1) we get that $\{\lambda \mid \mathrm{Re}\ \lambda \geq 0, |\lambda| \geq {}^p\sqrt{c+1}\}$ is contained in the resolvent set of A_p and for all λ in this set we have

$$\|(\lambda+A_p)^{-1}\| \leq \frac{c}{|\lambda|-c'}.$$

Now we deal with the set $|\lambda| \leq {}^p\sqrt{c+1}$. If $p \geq 2$ we can solve as before $(\lambda+A_p)u = f$ and get

$$\|u\|_{(H^{2,p}(\Omega))^n} \leq c(|\lambda| \cdot \|f\|_{(L^2(\Omega))^n} + \|f\|_{(L^p(\Omega))^n})$$

and therefore

$$\|(\lambda+A_p)^{-1}\| \le c, \quad \mathrm{Re}\ \lambda \ge 0.$$

Let $1 < p < 2$, $p \neq \frac{3}{2}$. For two solutions $u_1, u_2 \in D(A_p)$ of $(\lambda+A_p)u = f$ we get

$$u_2 - u_1 \in (H^{2,1}(\Omega))^n \subset (L^3(\Omega))^n$$

$$(\lambda+A_p)(u_2-u_1) = 0.$$

Using again the results of [L, 3.5] we see that $u_2-u_1 \in D(A_2)$ and therefore $u_2 = u_1$. Let $u_\nu \in D(A_p)$, $\nu = 1,2,\ldots$, be a sequence with

$$\|u_\nu\|_{H_p(\Omega)} = 1,$$

$$\|(\lambda+A_p)u_\nu\|_{H_p(\Omega)} \to 0.$$

Then the sequence $\{\|u_\nu\|_{(H^{2,p}(\Omega))^n}\}$ remains bounded (cf. III.1.10); thus we can choose a subsequence (which we denote by $\{u_\nu\}$ too) having the property that

$$u_\nu \to v \in D(A_p) \text{ weakly},$$

$$v \neq 0,$$

$$(\lambda+A_p)v = 0,$$

which is a contradiction. Thus

$$\text{(III.1.14)} \quad \|(\lambda+A_p)u\|_{H_p(\Omega)} \ge \delta(\lambda)\|u\|_{H_p(\Omega)}, \quad u \in D(A_p),$$

for some $\delta(\lambda) > 0$. Approximating again f in $H_p(\Omega)$ by $f_\nu \in C_0^\infty(\Omega)$ and denoting by u_ν the unique solution of $(\lambda+A_2)u = f_\nu$ we see that because of (III.1.8)

$$\|u_\nu - u_\mu\|_{(H^{2,p}(\Omega))^n} \leq c(\|f_\nu - f_\mu\|_{H_p(\Omega)} + |\lambda|\,\|u_\nu - u_\mu\|_{H_p(\Omega)}).$$

(III.1.14) gives the convergence of $\{u_\nu\}$ in $H_p(\Omega)$; thus $\{u_\nu\}$ converges in $(H^{2,p}(\Omega))^n$ against the unique solution $u \in D(A_p)$ of $(\lambda + A_p)u = f$. We have

$$\|u\|_{H^{2,p}(\Omega)} \leq c\|f\|_{L^p(\Omega)} + ({}^p\sqrt{c+1}/\delta)\|f\|_{L^p(\Omega)},$$

where ${}^p\sqrt{c+1}$ is the upper bound for $|\lambda|$. Thus $(\lambda + A_p)^{-1}$ exists in $\{\lambda \mid \mathrm{Re}\,\lambda \geq 0,\ |\lambda| \leq {}^p\sqrt{c+1}\}$. The last set is compact; therefore

$$\|(\lambda + A_p)^{-1}\| \leq c$$

on this set. Collecting all our estimates we arrive at

$$\|(\lambda + A_p)^{-1}\| \leq \frac{M'}{|\lambda|+1}, \quad \mathrm{Re}\,\lambda \geq 0.$$

The estimate (I.2.1) shows that A_p is of type

$$(\phi_{\varepsilon'}, \ \frac{1+\varepsilon'}{\varepsilon'} M' \sqrt{1 + (\frac{1}{M'(1+\varepsilon)})^2}$$

for all $\varepsilon' > 0$. Moreover in a set $\{\lambda \,\big|\, |\lambda| \leq \varepsilon'\}$ we have $\|(\lambda + A_p)^{-1}\| \leq \frac{M}{|\lambda|+1}$; the latter is furnished by the resolvent series in 0. Theorem III.1.3 is proved. □

<u>Remark:</u> 1. $c, \varepsilon, \Lambda_o$ <u>in theorem</u> III.1.3 <u>may depend on</u> p.

2. <u>Theorem</u> III.1.3 <u>together with</u> (I.2.1) <u>show that</u> $-A_p$ <u>generates an analytic semigroup</u> e^{-tA_p} <u>in</u> $H_p(\Omega)$; <u>moreover</u> e^{-tA_p} <u>decays with</u> $t \to \infty$, <u>more precisely we have</u>

(III.1.15) $\|e^{-tA_p}\| \leq ce^{-\delta t}$,

$\delta \in (0,1/M')$, as it was proved in I. § 1 and I. § 2.

3. A_p is called the Stokes-Operator.

4. Theorem III.1.1 is also valid for $p=\frac{3}{2}$ if $\varphi \in (H^{2,\frac{3}{2}}(\Omega) \cap \overset{o}{H}{}^{1,\frac{3}{2}}(\Omega))^n \cap H_{3/2}(\Omega)$, thus theorem III.1.3 remains valid for $p=\frac{3}{2}$ ([Sol 2]).

§ 2. Fractional Powers of the Stokes-Operator

As (III.1.15) shows we can introduce the fractional powers A_p^ζ for Re $\zeta > 0$ as it was done in I. § 3. We study here the basic properties of A_p^ζ, i.e. their domains of definition. By a simple proof we will also describe them not quite precisely but only "up to an $\varepsilon > 0$". It turns out that this is sufficient for some of our purposes. Ω, n are as in § 1.

One of our basic tools is the theorem of Hörmander-Mikhlin on Fourier multipliers, cf. [Hö, Tr2, p. 30]; using it we get some of our main results if Ω is starlike. Employing then a result of Heywood we get rid of the assumption "Ω starlike".

First we assume that Ω contains the point 0 and is starlike with respect to 0; this means that for every $\delta \in (0,1)$ the set $\Omega_\delta = \{\delta x \mid x \in \Omega\}$ fulfills the relation: $\overline{\Omega_\delta} \subset \Omega$.

Remark: In the sequel $\hat{\cdot}$ or F is used to denote the Fourier-Transform.

Theorem III.2.1: Let Ω be starlike with respect to 0. Then for all $s > 0$ and all $p > 1$

$$(\overset{o}{H}{}^{s,p}(\Omega))^n \cap H_p(\Omega) = \overline{\{\phi \mid \phi \in (C_0^\infty(\Omega))^n, \nabla\cdot\phi = 0\}}^{\|\cdot\|_{s,p}}.$$

Proof: The inclusion "$\supset$" is trivial. Thus we have to show the inclusion "$\subset$". For $r \in (0,1)$ we set

$$u_r(x) = \begin{cases} u(\frac{x}{r}), & x \in \Omega_r, \\ 0, & x \in \Omega - \Omega_r, \end{cases}$$

if u is any element from $(L^1(\Omega))^n$. The proof is divided into four steps.

Assertion 1: $u_r \in \overset{o}{H}{}^{s,p}(\Omega)$ if $u \in \overset{o}{H}{}^{s,p}(\Omega)$.

Proof: According to the definition of $\overset{o}{H}{}^{s,p}(\Omega)$ there is a sequence $\{\varphi^k\} \subset (C_o^\infty(\Omega))^n$ with

$$\|\varphi^k - u\|_{s,p} \to 0, \quad k \to \infty.$$

Let us consider φ_r^k. Evidently $\varphi_r^k \in (C_o^\infty(\Omega_r))^n$. For any $v \in L^2(\Omega)$ the formula

$$\hat{v}_r(\xi) = \frac{1}{(2\pi)^{n/2}} \int_{\mathbb{R}^n} e^{-i(\xi,x)} v(\tfrac{x}{r})\, dx,$$

$$(III.2.1) \quad = r^n \hat{v}(r\xi),$$

holds, where, of course, v is continued by 0 outside of Ω. Therefore for $\psi \in (C_o^\infty(\Omega))^n$, $\psi = (\psi_1, \ldots, \psi_n)$ we get

$$(III.2.2) \quad \|\psi_r\|_{s,p} \le c r^n \sum_{j=1}^n \left(\int_{\mathbb{R}^n} |F^{-1} \hat{\psi}_j(r\,.)(1+|\,.\,|^2)^{\frac{s}{2}}|^p \, d\tilde{\xi}\right)^{\frac{1}{p}},$$

$$\le c r^{n-\frac{n(p-1)}{p}} \sum_{j=1}^n \left(\int_{\mathbb{R}^n} |F^{-1}\hat{\psi}_j(\,.\,)(1+\frac{|\,.\,|^2}{r^2})^{\frac{s}{2}}|^p \, d\tilde{\xi}\right)^{\frac{1}{p}},$$

$$\le c r^{n-\frac{n(p-1)}{p}-s} \sum_{j=1}^n \left(\int_{\mathbb{R}^n} |F^{-1} \hat{\psi}_j(\,.\,)(r^2+|\,.\,|^2)^{\frac{s}{2}}|^p \, d\tilde{\xi}\right)^{\frac{1}{p}},$$

$$\leq c r^{n-\frac{n(p-1)}{p}-s} \sum_{j=1}^{n} \Big(\int_{\mathbb{R}^n} \Big| F^{-1} \hat{\psi}_j(.)(1+|.|^2)^{\frac{s}{2}} \cdot \frac{(r^2+|.|^2)^{\frac{s}{2}}}{(1+|.|^2)^{\frac{s}{2}}}\Big|^p d\tilde{\xi}\Big)^{\frac{1}{p}}.$$

It is easily seen that

$$\frac{1}{R^n} \int_{\frac{R}{2}\leq|\xi|\leq 2R} \left| R^{|\alpha|} D^\alpha \frac{(r^2+|\xi|^2)^{\frac{s}{2}}}{(1+|\xi|^2)^{\frac{s}{2}}}\right| d\xi \leq c(|\alpha|,s),$$

where $c(|\alpha|,s)$ does not depend on R. Then the theorem of Mikhlin-Hörmander (see [Tr2, p. 30]) shows that

$$\text{(III.2.3)} \quad \|\psi_r\|_{s,p} \leq c r^{n-n/p'-s} \sum_{j=1}^{n} \Big(\int_{\mathbb{R}^n} \Big| F^{-1} \hat{\psi}_j(.)(1+|.|^2)^{\frac{s}{2}} \cdot \frac{(r^2+|.|^2)^{\frac{s}{2}}}{(1+|.|^2)^{\frac{s}{2}}}\Big|^p d\tilde{\xi}\Big)^{\frac{1}{p}},$$

$$\leq c r^{n-n/p'-s} \sum_{j=1}^{n} \Big(\int_{\mathbb{R}^n} \Big| F^{-1} \hat{\psi}_j(.)(1+|.|^2)^{\frac{s}{2}}\Big|^p d\tilde{\xi}\Big)^{\frac{1}{p}},$$

$$\leq c r^{n-n/p'-s} \|\psi\|_{s,p},$$

where $p' = p/(p-1)$. Therefore we get that $\|\varphi_r^k - \varphi_r^l\|_{s,p} \to 0$ for $k,l \to \infty$, i.e. there is a $v_r \in (\overset{\circ}{H}{}^{s,p}(\Omega))^n$ with $\varphi_r^k \to v_r$ in $(\overset{\circ}{H}{}^{s,p}(\Omega))^n$. On the other hand we have $u_r \in (L^p(\Omega))^n$ and $\|u_r - \varphi_r^k\|_{(L^p(\Omega))^n} \to 0$. Therefore $v_r = u_r$ and $u_r \in (\overset{\circ}{H}{}^{s,p}(\Omega))^n$.

<u>Assertion 2: Let</u> $u \in (\overset{o}{H}{}^{s,p}(\Omega))^n$. <u>Then</u>

$$\|u-u_r\|_{s,p} \to 0 \text{ for } r \to 1.$$

<u>Proof:</u> Let $\varepsilon > 0$. Then there is a $\varphi \in (C_o^\infty(\Omega))^n$ with

$$\|u-\varphi\|_{s,p} \leq \frac{\varepsilon}{3}.$$

We have

$$\|u-u_r\|_{s,p} \leq \|u-\varphi\|_{s,p} + \|\varphi-\varphi_r\|_{s,p} + \|\varphi_r-u_r\|_{s,p}.$$

According to (III.2.3) we get

$$\|\varphi_r-n_r\|_{s,p} \leq cr^{n-n/p'-s}\|\varphi-u\|_{s,p'}$$
$$\leq cr^{n-n/p'-s}\cdot\frac{\varepsilon}{3}.$$

Since $\varphi \in (C_o^\infty(\Omega))^n$ it follows that $\hat{\varphi}$ can be estimated in the following way: There exists a $c(\varphi)$ such that

$$|\hat{\varphi}(\xi)| \leq \frac{c(\varphi)}{(1+|\xi|^2)^{\frac{n+1+s}{2p'}}}.$$

Then the equality (III.2.1) shows that

$$|\hat{\varphi_r}(\xi)| \leq r^n|\hat{\varphi}(r\xi)|$$

$$\leq \frac{c(\varphi)r^n}{(1+|r\xi|^2)^{\frac{n+1+s}{2p'}}}$$

$$\leq \frac{c(\varphi)}{(1+|\xi|^2)^{\frac{n+1+s}{2p'}}},\ \frac{1}{2} \leq r \leq 1,$$

$$|\hat{\varphi}(\xi)-\hat{\varphi}_r(\xi)| \leq \frac{c(\varphi)}{(1+|\xi|^2)^{\frac{n+1+s}{2p'}}},$$

$$|\hat{\varphi}(\xi)-\hat{\varphi}_r(\xi)|(1+|\xi|^2)^{\frac{s}{2}} \leq \frac{c(\varphi)}{(1+|\xi|^2)^{\frac{n+1}{2p'}}}.$$

Since we have

$$\|\varphi-\varphi_r\|_{s,p} \leq c\Big(\int_{\mathbb{R}^n} \sum_{j=1}^{n} |\hat{\varphi}_j(\xi)-\hat{\varphi}_{jr}(\xi)|^{p'}(1+|\xi|^2)^{\frac{sp'}{2}}\, d\xi\Big)^{\frac{1}{p'}}.$$

By Lebesgue 's theorem of dominated convergence we get $\|\varphi-\varphi_r\|_{s,p} \to 0$ for $r \to 1$. For r, $r_o(\varepsilon) \leq r \leq 1$ we therefore obtain

$$\|\varphi-\varphi_r\|_{s,p} \leq \frac{\varepsilon}{3}.$$

This yields assertion 2.

Now let $\Omega' \subset\subset \Omega$, Ω' open, $v \in (\overset{o}{H}{}^{s,p}(\Omega))^n$, $v(x) = 0$ on $\Omega-\Omega'$. Let $\varphi \in (C_o^\infty(\Omega))^n$, $\varphi \geq 0$, $\int_{\mathbb{R}^n} \varphi(x)dx = 1$, $\varphi_\varepsilon(x) = \varphi(\frac{x}{\varepsilon})$ and

$$v_\varepsilon(x) = \int_{\mathbb{R}^n} \varphi_\varepsilon(x-y)v(y)\, dy, \ \varepsilon > 0.$$

<u>Assertion 3:</u> <u>For</u> $0 < \varepsilon < \operatorname{dist}(\Omega', \partial\Omega)$ <u>we have</u>

$$v_\varepsilon \in (C_o^\infty(\Omega))^n,$$

$$\|v-v_\varepsilon\|_{s,p} \to 0 \text{ for } \varepsilon \to 0.$$

<u>Proof:</u> It is clear that $v_\varepsilon \in (C_o^\infty(\Omega))^n$. The definition of v_ε furnishes

$$\hat{v}_\varepsilon = (2\pi)^{n/2}\varphi_\varepsilon \cdot \hat{v}.$$

Since

$$\varphi_\varepsilon^\wedge(\xi) = \frac{1}{(2\pi)^{n/2}} \int_{\mathbb{R}^n} e^{i(\xi,x)} \varepsilon^{-n} \varphi(\frac{x}{\varepsilon})\, dx = \hat{\varphi}(\varepsilon\xi)$$

and consequently

$$\| v - v_\varepsilon \|_{s,p} = \| F^{-1} \sum_{j=1}^{n} (\hat{v}_j(.) - \hat{v}_{j\varepsilon}(.))(1+|.|^2)^{\frac{s}{2}} \|_{L^p(\Omega)}$$

$$= \| F^{-1} \sum_{j=1}^{n} (\hat{v}_j(.)(1-(2\pi)^{n/2}\hat{\varphi}(\varepsilon .))(1+|.|^2)^{\frac{s}{2}} \|_{L^p(\mathbb{R}^n)}.$$

Let us apply this relation if $v = \psi \in (C_o^\infty(\Omega))^n$. If we use again the theorem of Mikhlin-Hörmander (see also [Tr2, p. 30]) of our auxiliary propositions we get for a suitable $N \in \mathbb{N}$

$$\| \psi - \psi_\varepsilon \|_{s,p} \le c \cdot \sup_{|\alpha| \le N} \sup_{\substack{\xi \in \mathbb{R}^n, \\ 1 \le j \le n}} (1+|\xi|^2)^{\frac{|\alpha|+n+1+s}{2}} |D^\alpha \hat{\psi}_j(\xi)| \cdot$$

$$\cdot \sum_{j=1}^{n} \| F^{-1} \frac{1-\hat{\varphi}(\varepsilon .)}{(|.|^2+1)^{\frac{n+1+s}{2}}} \cdot (1+|.|^2)^{\frac{s}{2}} \|_{L^p(\mathbb{R}^n)},$$

$$\le c \cdot \sup_{|\alpha| \le N} \sup_{\substack{\xi \in \mathbb{R}^n \\ 1 \le j \le n}} (1+|\xi|^2)^{\frac{|\alpha|+n+1}{2}} \cdot |D^\alpha \hat{\psi}_j(\xi)| \cdot$$

$$\cdot \sum_{j=1}^{n} \| F^{-1} \frac{1}{(|.|^2+1)^{\frac{n+1}{2}}} - \varphi_\varepsilon * F^{-1} \frac{1}{(|.|^2+1)^{\frac{n+1}{2}}} \|_{L^p(\mathbb{R}^n)};$$

observe that since $\frac{1}{(1+|\xi|^2)^{\frac{n+1}{2}}} \in \bigcap_{1 \le q \le 2} L^q(\mathbb{R}^n)$ we have that

$F^{-1} \frac{1}{(1+|.|^2)^{\frac{n+1}{2}}} \in L^p(\mathbb{R}^n)$. Thus the last sum is converging to 0, $\varepsilon \to 0$, and we get

(III.2.4) $\|\psi-\psi_\varepsilon\|_{s,p} \to 0$

for $\varepsilon \to 0$. For an arbitrary $\eta > 0$ we choose a $\psi \in (C_0^\infty(\Omega))^n$ with

$$\|v-\psi\|_{s,p} < \frac{\eta}{3}.$$

Then we have

$$\begin{aligned}\|(v-\psi)_\varepsilon\|_{s,p} &= \sum_{j=1}^{n} \|F^{-1}(\hat{v}_j(.)-\hat{\psi}_j(.))(2\pi)^{n/2}\hat{\varphi}(\varepsilon .) \cdot (1+|.|^2)^{\frac{s}{2}}\|_{L^p(\mathbb{R}^n)},\\ &\le c \sup_{|\alpha|\le N} \sup_{\xi\in\mathbb{R}^n} (1+|\xi|^2)^{\frac{|\alpha|}{2}} |D^\alpha\hat{\varphi}(\varepsilon\xi)| \\ &\qquad \sum_{j=1}^{n} \|F^{-1}(\hat{v}_j(.)-\hat{\psi}_j(.))(1+|.|^2)^{\frac{s}{2}}\|_{L^p(\mathbb{R}^n)},\\ &\le c \sup_{|\alpha|\le N} \sup_{\xi\in\mathbb{R}^n} (1+|\xi|^2)^{\frac{|\alpha|}{2}} |D^\alpha\hat{\varphi}(\varepsilon\xi)|\, \|v-\psi\|_{s,p},\end{aligned}$$

with $N \in \mathbb{N}$ as above. Since

$$|D^\alpha\hat{\varphi}(\varepsilon\xi)| \le \varepsilon^{|\alpha|} \cdot \frac{c(\varphi)}{|\xi|^{N}+1}$$

we get

$$\|(v-\psi)_\varepsilon\|_{s,p} \le c(\varphi)\|v-\psi\|_{s,p}.$$

Thus we can assume that also

$$\|(v-\psi)_\varepsilon\|_{s,p} \le \frac{\eta}{3}, \quad 0 < \varepsilon \le 1.$$

Collecting our inequalities we get

$$\|v-v_\varepsilon\|_{s,p} \leq \|v-\psi\|_{s,p} + \|\psi-v_\varepsilon\|_{s,p},$$
$$\leq \|v-\psi\|_{s,p} + \|\psi-\psi_\varepsilon\|_{s,p} + \|(\psi-v)_\varepsilon\|_{s,p}$$

and consequently assertion 3.

<u>Assertion 4:</u> <u>Let</u> $u \in (\overset{\circ}{H}{}^{s,p}(\Omega))^n \cap H_p(\Omega)$, $0 < r < 1$, $0 < \varepsilon < 1-r$. <u>Then</u>

$$u_{r\varepsilon} \in (C_o^\infty(\Omega))^n,$$
$$\nabla \cdot u_{r\varepsilon} = 0.$$

<u>Proof:</u> Obviously $u_{r\varepsilon} \in (C_o^\infty(\Omega))^n$. We have

$$\nabla \cdot u_{r\varepsilon} = \nabla \cdot (\varphi_\varepsilon * u_r).$$

There is a sequence $\{\varphi^k\}$ with $\varphi^k \in (C_o^\infty(\Omega))^n$, $\nabla \cdot \varphi^k = 0$ and $\|u-\varphi^k\|_{(L^p(\Omega))^n} \to 0$, $k \to \infty$. Therefore

$$\|u_r-\varphi_r^k\|_{(L^p(\Omega))^n} \to 0, \ k \to \infty$$

and

$$\nabla \cdot \varphi_r^k(x) = \frac{1}{r}\nabla \cdot \varphi^k\left(\frac{x}{r}\right) = 0.$$

We obtain

$$|\nabla \cdot u_{r\varepsilon}| \leq |\nabla \cdot (\varphi_\varepsilon * (u_r-\varphi_r^k)| + |\nabla \cdot (\varphi_\varepsilon * \varphi_r^k)),$$
$$= |\nabla \cdot (\varphi_\varepsilon * (u_r-\varphi_r^k))|,$$
$$\leq \|\sum_{i=1}^n \frac{\partial}{\partial x_i}\varphi_\varepsilon\|_{L^2(\Omega)} \|u_r-\varphi_r^k\|_{L^2(\Omega)}.$$

Letting k tend to $+\infty$ we get the assertion. Now we proceed with the <u>proof of our original theorem</u>: We choose a sequence $\{r_k\}$ with $r_k \in (0,1)$, $r_k \to 1$, $k \to \infty$. For every k we fix an ε_k such that $0 < \varepsilon_k < 1-r_k$ and

$$\|u_{r_k} - u_{r_k\varepsilon_k}\|_{s,p} \leq \frac{1}{k}$$

which is possible according to assertion 3. Let $\Theta_k = u_{r_k\varepsilon_k}$. Then $\Theta_k \in (C_o^\infty(\Omega))^n$, $\nabla\cdot\Theta_k = 0$ according to assertion 4, and

$$\|u-\Theta_k\|_{s,p} \leq \|u-u_{r_k}\|_{s,p} + \|u_{r_k} - u_{r_k\varepsilon_k}\|_{s,p'}$$

$$\leq \|u-u_{r_k}\|_{s,p} + \frac{1}{k}.$$

Assertion 2 furnishes

$$\|u-u_{r_k}\|_{s,p} \to 0, \quad k \to \infty.$$

theorem III.2.1 is proved. □

<u>Lemma III.2.1:</u> <u>Let</u> Ω <u>be as in theorem</u> III.2.1. <u>Then</u>

$$(\overset{o}{H}{}^{1,p}(\Omega))^n \cap H_p(\Omega) = (\overset{o}{H}{}^{1,p}(\Omega))^n \cap \{u \mid u \in (H^{1,p}(\Omega))^n, \ \nabla\cdot u = 0\},$$

$p \geq 2$.

<u>Proof:</u> Let $u \in (\overset{o}{H}{}^{1,p}(\Omega))^n$, $\nabla\cdot u = 0$. We can construct $u_{r\varepsilon} \in (C_o^\infty(\Omega))^n$ as in the proof of assertion 4 before and get that $\nabla\cdot u_{r\varepsilon} = 0$. Then it can be shown as in the proof of theorem III.2.1 that $u \in \overline{\{\psi \mid \psi \in (C_o^\infty(\Omega))^n, \ \nabla\cdot\psi = 0\}}^{\|\cdot\|_{1,p}}$ and consequently $u \in (\overset{o}{H}{}^{1,p}(\Omega))^n \cap$ $\cap H_p(\Omega)$. On the other hand it follows from assertion 4 that any u from $(\overset{o}{H}{}^{1,p}(\Omega))^n \cap H_p(\Omega)$ has the property that $\nabla\cdot u = 0$. Our lemma is proved. □

Now we want to get rid from the assumption that Ω is starlike. In this case the proof of theorem III.2.1 essentially is due to Heywood [He]. We cannot give the complete proof here but want to exhibit the major steps in Heywood's proof. According to [He, pp. 78, 81] for every r, $0<r\leq 1$, there exists a mapping $T_r:\Omega\to\Omega$ of class C^3 with the following properties:

1. T_1 is the identity
2. T_r is unique for every $r\in(0,1]$, $\overline{T_r(\Omega)}\subset\Omega$, $r\in(0,1)$.
3. $T_r, \nabla T_r, \nabla^2 T_r, \nabla^3 T_r$ are uniformly continuous in dependence of $(r,x)\in(0,1]\times\Omega$.

We set

$$u_r(x) = \begin{cases} \dfrac{\nabla T_r(x')\cdot u(x')}{J_r(x')}, & x = T_r(x')\in T_r(\Omega), \\ 0, & x\in\Omega-T_r(\Omega), \end{cases}$$

where $J_r(x)$ denotes the Jacobian of T_r in x. By $\nabla T_r(x')\cdot u(x')$ we mean the vector with components

$$\sum_{i=1}^{n} \frac{\partial T_{rl}(x')}{\partial x_i}\cdot u_i(x'),\ 1\leq l\leq n,$$

where T_{rl}, $1\leq l\leq n$, are the components of T_r. For some $\delta,\varepsilon\in(0,1)$ we have

$$J_r(x)\geq\varepsilon>0,\ r\in[\delta,1].$$

The inverse S_r of T_r is of class C^2 and $\nabla S_r, \nabla^2 S_r$ are uniformly continuous in dependence of $(r,x)\in[\delta,1]\times T_r(\Omega)$. Moreover, for $r\in[\delta,1]$ we have $u_r\in(C_o^1(\Omega))^n$ if $u\in(C_o^1(\Omega))^n$. For a proof of these assertions see [He, p. 79]. If $u\in(\overset{o}{H}{}^{2,p}(\Omega))^n$ then $u_r\in(\overset{o}{H}{}^{2,p}(\Omega))^n$, $r\in[\delta,1]$ and

$$\|u_r\|_{2,p}\leq\tilde{c}\|u\|_{2,p};$$

with $u \in (L^p(\Omega))^n$ we get that $u_r \in (L^p(\Omega))^n$, $r \in [\delta,1]$, and

$$\|u_r\|_{(L^p(\Omega))^n} \le \tilde{c}\|u\|_{(L^p(\Omega))^n}$$

([He, p. 79]). Therefore the mapping

$$A_r : u \to u_r$$

is boundedly linear from $(\overset{\circ}{H}{}^{2,p}(\Omega))^n$ into itself and from $(L^p(\Omega))^n$ into itself. Since $(\overset{\circ}{H}{}^{s,p}(\Omega))^n$ is the complex interpolation space $[(L^p(\Omega))^n, (\overset{\circ}{H}{}^{2,p}(\Omega))^n]_{s/2}$, $0 \le s \le 2$, $s \neq \frac{1}{p}$, $s \neq 1+\frac{1}{p}$, A_r is boundedly linear from $(\overset{\circ}{H}{}^{s,p}(\Omega))^n$ into itself with bound $\tilde{c}$, $s \neq \frac{1}{p}$, $s \neq 1+\frac{1}{p}$. This is the analogue to assertion 1 in the proof of theorem III.2.1. Also $\|u_r - u\|_{2,p} \to 0$ for $r \to 1$ and in particular $\|\varphi_r - \varphi\|_{s,p} \to 0$ for $r \to 1$, $0 \le s \le 2$, $s \neq \frac{1}{p}$, $s \neq 1+\frac{1}{p}$, $\varphi \in (C_o^\infty(\Omega))^n$. This gives the analogue to assertion 2 in the proof of theorem III.2.1. Of course assertion 3 remains still valid, and the first part of assertion 4 is obvious; the second part, namely $\nabla \cdot u_{r\varepsilon} = 0$, can be proved as in the proof of assertion 4 (cf. [He, p. 80]). Now we continue as in the proof of theorem III.2.1 and lemma III.2.1 to get

<u>Theorem III.2.2:</u> <u>Let</u> Ω <u>be as in</u> III.1. <u>Let</u> $p \ge 2$, $0 \le s \le 2$, $s \neq \frac{1}{p}$, $s \neq 1+\frac{1}{p}$. <u>Then we get the relations</u>

$$(\overset{\circ}{H}{}^{s,p}(\Omega))^n \cap H_p(\Omega) = \overline{\{\psi \mid \psi \in (C_o^\infty(\Omega))^n,\ \nabla \cdot \psi = 0\}}^{\|\cdot\|_{s,p}},$$

$$(\overset{\circ}{H}{}^{1,p}(\Omega))^n \cap H_p(\Omega) = \{u \mid u \in (\overset{\circ}{H}{}^{1,p}(\Omega))^n,\ \nabla \cdot u = 0\}.$$

The restriction $s \neq \frac{1}{p}$, $s \neq 1+\frac{1}{p}$ is not important for our purposes and can be dropped if $(\overset{\circ}{H}{}^{s,p}(\Omega))^n$ denotes the interpolation space $[(L^p(\Omega))^n, (\overset{\circ}{H}{}^{2,p}(\Omega))^n]_{s/2}$, $0 \leq s \leq 2$. Besides the complex interpolation space we need the real interpolation space

$$((L^p(\Omega))^n, (\overset{\circ}{H}{}^{2,p}(\Omega))^n)_{s/2,p}, \quad 0 \leq s \leq 2$$

(cf. the auxiliary propositions 0.2). We have

$$(\overset{\circ}{W}{}^{s,p}(\Omega))^n = ((L^p(\Omega))^n, (\overset{\circ}{H}{}^{2,p}(\Omega))^n)_{s/2,p}, \quad 0 \leq s \leq 2, \quad s \neq \frac{1}{p}, \neq 1+\frac{1}{p}$$

(cf. the auxiliary propositions 0.2). Using the real interpolation spaces we can prove the following theorem:

<u>Theorem III.2.3:</u> <u>We have</u>

$$(\overset{\circ}{H}{}^{2s,p}(\Omega))^n \cap H_p(\Omega) \subset D(A_p^\gamma), \quad 0 \leq \gamma < s,$$

<u>algebraically and topologically if</u> $0 \leq s \leq 1$, $p \geq 2$, $2s \neq \frac{1}{p}$, $2s \neq \frac{1}{p}+1$.

<u>Proof:</u> Since algebraically and topologically

$$(\overset{\circ}{W}{}^{2s+\varepsilon,p}(\Omega))^n \subset (\overset{\circ}{H}{}^{2s,p}(\Omega))^n \subset (\overset{\circ}{W}{}^{2s-\varepsilon,p}(\Omega)), \quad 0 < \varepsilon \leq 2s$$

(cf. the auxiliary propositions, (0.2.13)) we can prove the theorem in question with $(\overset{\circ}{W}{}^{2s,p}(\Omega))^n$ instead of $(\overset{\circ}{H}{}^{2s,p}(\Omega))^n$. The proof is essentially due to Kielhöfer [Ki2, p. 136]. We use a representation formula from [BBW] for the fractional powers A_p^γ, $0 < \gamma < 1$. Namely, $u \in D(A^\gamma)$ if and only if

$$\int_0^\infty (e^{-sA_p} - I)s^{-\gamma-1}u\, ds$$

in $H_p(\Omega)$ exists; then we have

$$A^\gamma u = \frac{1}{\Gamma(1-\gamma)} \int_0^\infty (e^{-sA_p} - I)s^{-\gamma-1}u\, ds.$$

According to [BB, p. 194] the real interpolation space $(H_p(\Omega),D(A_p))_{\theta,p}$ is exactly the set

$$(H_p(\Omega),D(A_p))_{\theta,p} = \left\{u \mid \|u\|_{H_p(\Omega)} + \left(\int_0^\infty (s^{-\theta}\|(e^{-sA_p}-I)u\|_{H_p(\Omega)})^p\frac{ds}{s}\right)^{\frac{1}{p}} < +\infty\right\},\ 0<\theta<1,$$

where $D(A_p)$ is endowed by the norm $\|.\|_{2,p}$. The term

$$\left(\int_0^\infty (s^{-\theta}\|(e^{-sA_p}-I)u\|_{H_p(\Omega)})^p \frac{ds}{s}\right)^{\frac{1}{p}} + \|u\|_{H_p(\Omega)}$$

defines a norm in $(H_p(\Omega),D(A_p))_{\theta,p}$. We then have

$$\begin{aligned}
\|A^\gamma u\|_{H_p(\Omega)} &\le \frac{1}{\Gamma(1-\gamma)}\int_0^\infty \|(e^{-sA_p}-I)u\|s^{-\gamma-1}\,ds,\\
&\le c\left(\int_0^\infty \|(e^{-sA_p}-I)u\|s^{-\gamma-1}\,ds + \|u\|_{H_p(\Omega)}\right),\\
&\le c(\theta)\left\{\left(\int_0^1 \|(e^{-sA_p}-I)u\|^p s^{-p\theta}\frac{ds}{s}\right)^{\frac{1}{p}} + \|u\|_{H_p(\Omega)}\right\},\\
&\le c(\theta)\|u\|_{(H_p(\Omega),D(A_p))_{\theta,p}},\quad 1>\theta>\gamma.
\end{aligned}$$

Since $(\overset{\circ}{H}{}^{2,p}(\Omega)^n \cap H_p(\Omega)) \subset D(A_p)$ with a continuous imbedding we have

$$(H_p(\Omega),(\overset{\circ}{H}{}^{2,p}(\Omega))^n \cap H_p(\Omega))_{\theta,p} \subset (H_p(\Omega),D(A_p))_{\theta,p}$$

with a continuous imbedding (cf. our auxiliary propositions 0.1). Our theorem is proved. □

<u>Corollary to theorem III.2.3:</u> <u>Let</u> $0 \le 2s < \frac{1}{p}$. <u>Then</u>

$$(H^{2s,p}(\Omega))^n \cap H_p(\Omega) = (\overset{\circ}{H}{}^{2s,p}(\Omega))^n \cap H_p(\Omega) \subset D(A^\gamma),\ 0 \le \gamma < s.$$

<u>Proof:</u> The first equality follows from our auxiliary propositions 0.2. The last inclusion is a consequence of theorem III.2.3. □

We also need somewhat an opposite direction of theorem III.2.3, namely

<u>Theorem III.2.4:</u> <u>We have algebraically and topologically</u>

$$D(A^{\gamma}) \subset (H^{2s,p}(\Omega))^n \cap H_p(\Omega), \; 1 \geq \gamma > s.$$

<u>If</u> $p > n$ <u>then</u>

$$(C^{1+\alpha}(\bar{\Omega}))^n \supset D(A^{\gamma})$$

<u>if</u> $\alpha < 1-\frac{n}{p}$, $1 \geq \gamma > \frac{1}{2}+\frac{n}{2p}+\frac{\alpha}{2}$, <u>and the estimate</u>

$$\text{(III.2.5)} \quad \|u\|_{(C^{1+\alpha}(\bar{\Omega}))^n} \leq c\|A^{\gamma}u\|_{H_p(\Omega)}$$

<u>holds</u>.

<u>Proof:</u> Using Sobolev's imbedding in 0.2:

$$\|u\|_{(C^{1+\alpha}(\bar{\Omega}))^n} \leq c\|u\|_{2s,p}, \; p > n, \; \alpha < 1-\frac{n}{p}, \; 2 > 2s > 1+\alpha+\frac{n}{p},$$

we get by the interpolation inequality the estimate

$$\|u\|_{(C^{1+\alpha}(\bar{\Omega}))^n} \leq c\|u\|_{2,p}^{\frac{2s}{2}}\|u\|_{H_p(\Omega)}^{1-\frac{2s}{2}}.$$

We have already proved that

$$\|u\|_{2,p} \leq c\|A_pu\|_{H_p(\Omega)};$$

Consequently

$$\|u\|_{(C^{1+\alpha}(\bar{\Omega}))^n} \leq c\|A_pu\|^s\|u\|_{H_p(\Omega)}^{1-s}.$$

The next step is carried through as in [F, pp. 177,178]. It follows that

$$\text{(III.2.6)}\quad \|u\|_{(C^{1+\alpha}(\bar{\Omega}))^n} \leq c_1\|A_p u\|\delta^{1-s} + c_2\|u\|_{H_p(\Omega)}\delta^{-s},$$

$\delta > 0$. Since $\gamma > s$ we get

$$\|A^{-\gamma}u\|_{(C^{1+\alpha}(\bar{\Omega}))^n} \leq \frac{1}{\Gamma(\gamma)}\int_0^1 \|e^{-tA_p}u\|_{(C^{1+\alpha}(\bar{\Omega}))^n} t^{\gamma-1}\,dt + \frac{1}{\Gamma(\gamma)}\int_1^{+\infty} \|e^{-tA_p}u\|_{(C^{1+\alpha}(\bar{\Omega}))^n} t^{\gamma-1}\,dt.$$

Using (III.2.6) with $\delta = t$ we get

$$\|e^{-tA_p}u\|_{(C^{1+\alpha}(\bar{\Omega}))^n} \leq c_1\|A_p e^{-tA_p}u\|_{H_p(\Omega)} t^{1-s} + c_2\|e^{-tA_p}u\|_{H_p(\Omega)} t^{-s},$$

$$\int_0^1 \|e^{-tA_p}u\|_{(C^{1+\alpha}(\bar{\Omega}))^n} t^{\gamma-1}\,dt \leq c\int_0^1 t^{\gamma-s-1}\,dt\,\|u\|_{H_p(\Omega)} \leq c\|u\|_{H_p(\Omega)}.$$

Using (III.2.6) with $\delta = 1$ and the exponential decay of e^{-tA_p} we get

$$\int_1^{+\infty} \|e^{-tA_p}u\|_{(C^{1+\alpha}(\bar{\Omega}))^n} t^{\gamma-1}\,dt \leq c\int_1^{\infty} e^{-\tilde{\delta}t}\,dt\,\|u\|_{H_p(\Omega)}$$

for some $\tilde{\delta} > 0$. Collecting our inequalities we see that

$$\|A_p^{-\gamma}u\|_{C^{1+\alpha}(\bar{\Omega})} \leq c\|u\|_{H_p(\Omega)}$$

and our theorem is proved. □

As we see from theorems III.2.3, III.2.4 our description of the domains of A_p^γ is precise only up to an $\varepsilon > 0$. A first improvement can be given by means of the Heinz inequality in the case p = 2 ([Ta, p. 44]): Let X_1, X_2 be two Hilbert spaces and let A,B be two selfadjoint positive operators in X_1, X_2 resp. Let T be a bounded operator from X_1 into X_2 which maps D(A) into D(B). Let

$$\|BTu\|_{X_1} \leq M\|Au\|_{X_2}, \quad u \in D(A),$$

for some $M > 0$. Then

(III.2.7) $TD(A^\alpha) \subset D(B^\alpha), \quad 0 < \alpha < 1,$

and

(III.2.8) $\|B^\alpha Tu\|_{X_1} \leq M^\alpha \|T\|^{1-\alpha} \|A^\alpha u\|_{X_2},$

$u \in D(A^\alpha)$.

Theorem III.2.5: Let T *be the identity from* $H_2(\Omega)$ *into* $(L^2(\Omega))^n$. *Let* $A = A_2$ *the positive selfadjoint operator* $-\nu P\Delta$ *in* $H_2(\Omega)$ *with domain* $D(A) = D(A_2) = (H^{2,2}(\Omega) \cap \overset{\circ}{H}{}^{1,2}(\Omega))^n \cap H_2(\Omega)$. *Let* B *be the positive selfadjoint operator* $-\nu\Delta$ *in* $(L^2(\Omega))^n$ *with domain of definition* $(H^{2,2}(\Omega) \cap \overset{\circ}{H}{}^{1,2}(\Omega))^n$. *Then* $D(A^\gamma) \subset D(B^\gamma)$

(III.2.9) $\|B^\gamma u\|_{(L^2(\Omega))^n} \leq \|A^\gamma u\|_{H_2(\Omega)}, \quad 0 \leq \gamma \leq 1, \quad u \in D(A_2)$

Moreover $D(B^\gamma) \subset (H^{2\gamma,2}(\Omega))^n$, $1 \leq \gamma \leq 1$, $D(B^\gamma) = (\overset{\circ}{H}{}^{2\gamma,2}(\Omega))^n$, $0 \leq \gamma \leq \frac{1}{2}$, $\gamma \neq \frac{1}{4}$,

(III.2.10) $\frac{1}{c}\|u\|_{2\gamma,2} \leq \|B^\gamma u\|_{H_2(\Omega)}, \quad 0 \leq \gamma \leq 1,$

(III.2.11) $\|B^\gamma u\|_{H_2(\Omega)} \leq c\|u\|_{2\gamma,2}, \quad 0 \leq \gamma \leq 1, \quad \gamma \neq \frac{1}{4}.$

Proof: (III.2.9) was proved in [FK] and readily follows from (III.2.7) and (III.2.8). (III.2.10) and (III.2.11) are proved in [Fu] and [W3, Tr1]. □

The generalization of theorem III.2.5 to arbitrary integration exponents $p>1$ is due to Giga [Gi1, Gi2]. The proof of this important result is based on the estimate

$$\text{(III.2.12)}\quad \|A_p^{i\tau}u\|_{H_p(\Omega)} \le ce^{c|\tau|}\|u\|_{H_p(\Omega)},\quad \tau \in \mathbb{R}.$$

Once having proved this estimate one can show that $D(A_p^\gamma)$ is the complex interpolation space $[H_p(\Omega),D(A_p)]_\gamma$ and compare $D(A_p^\gamma)$ with $D(B_p^\gamma) = [(L^p(\Omega))^n,D(B_p)]_\gamma$ ($B_p = -\nu\Delta$ in $(L^p(\Omega))^n$ with domain of definition $(H^{2,p}(\Omega)\cap \overset{\circ}{H}{}^{1,p}(\Omega))^n$, $D(A_p)$ and $D(B_p)$ are normed by $\|\cdot\|_{2,p}$; the preceding characterisation of $D(B_p^\gamma)$ is also due to Fujiwara [Fu] and based on an estimate for $B_p^{i\tau}$ like (III.2.12)). We cannot give a proof of Giga's theorem here; we will only formulate it and apply it sometimes in what follows.

Theorem III.2.6: Let $p>1$. We have

$$\text{(III.2.13)}\quad D(A_p^\gamma) = D(B_p^\gamma)\cap H_p(\Omega),\quad 0\le\gamma\le 1,\quad p>1$$

and

$$\text{(III.2.14)}\quad \frac{1}{c}\|B_p^\gamma u\|_{(L^p(\Omega))^n} \le \|A_p^\gamma u\|_{H_p(\Omega)} \le c\|B_p^\gamma u\|_{(L^p(\Omega))^n}.$$

Moreover

$$D(A_p^\gamma) \subset (H^{2\gamma,p}(\Omega))^n\cap H_p(\Omega),\quad 0\le\gamma\le 1,$$

$$D(A_p^\gamma) = (\overset{\circ}{H}{}^{2\gamma,p}(\Omega))^n\cap H_p(\Omega),\quad 0\le\gamma\le\frac{1}{2},\quad \gamma \ne \frac{1}{2p},$$

$$\frac{1}{c}\|u\|_{2\gamma,p} \leq \|A_p^{\gamma}u\|_{H_p(\Omega)}, \quad 0 \leq \gamma \leq 1, \quad u \in D(A_p^{\gamma}),$$

$$\|A_p^{\gamma}u\|_{H_p(\Omega)} \leq c\|u\|_{2\gamma,p}, \quad 0 \leq \gamma \leq 1, \quad \gamma \neq \frac{1}{2p}, \quad u \in D(A_p^{\gamma}).$$

An important consequence of (III.2.12) is an estimate for $\|A_p^{-\sigma}u\|$, $0 \leq \sigma < \frac{1}{2p}$, namely

<u>Theorem III.2.7:</u> <u>Let</u> $p > 1$, $u \in H_p(\Omega)$. <u>Then</u>

$$\|B_p^{-\sigma}u\|_{(L^p(\Omega))^n} \leq c(\varepsilon)\|A_p^{-\sigma}u\|_{H_p(\Omega)},$$

$0 \leq \sigma \leq \frac{1}{2p}-\varepsilon$, $u \in H_p(\Omega)$, $0 < \varepsilon \leq \frac{1}{2p}$.

<u>Proof:</u> We use Hadamard's Three Lines Theorem again. Let $\zeta = \sigma+i\tau$, $0 \leq \sigma \leq 1$, $\tau \in \mathbb{R}$. According to Giga [Gi] and Fujiwara [Fu] resp. we have

$$\|A_p^{i\tau}\| \leq \tilde{c}e^{\tilde{c}|\tau|},$$

$$\|B_p^{i\tau}\| \leq \tilde{c}e^{\tilde{c}|\tau|},$$

$\tau \in \mathbb{R}$, for some constant $\tilde{c} > 0$. Then for $u \in D(A_p)$, $v \in L^p(\Omega)$ we consider for some $\varepsilon \in (0,\frac{1}{2p}]$ the function

$$f(\zeta) = \frac{1}{\tilde{c}}\langle A_p^{(\frac{1}{2p}-\varepsilon)\zeta}u, B_q^{-(\frac{1}{2p}-\varepsilon)\overline{\zeta}}v\rangle e^{-\tilde{c}(\frac{1}{2p}-\varepsilon)^2\zeta^2}$$

which is bounded in $0 \leq \sigma \leq 1$, holomorphic in $0 < \sigma < 1$; furthermore $\langle.,.\rangle$ is the L^p-L^q-duality. We have the estimates

$$|f(0+i\tau)| \leq \|u\|_{H_p(\Omega)}\|v\|_{L^q(\Omega)}.$$

From theorem III.2.6 it follows that

$$|f(1+i\varepsilon)| = |\frac{1}{\tilde{c}}\langle A_p^{\varepsilon\zeta}u, B_p^{-\varepsilon\bar{\zeta}}v\rangle e^{-\varepsilon^2\tilde{c}\zeta^2}|,$$

$$(\text{III.2.15}) \qquad = |\frac{1}{\tilde{c}}\langle A_p^{\varepsilon\zeta}u, P_pB_p^{-\varepsilon\bar{\zeta}}v\rangle e^{-\varepsilon^2\tilde{c}\zeta^2}|,$$

$$(\text{III.2.16}) \qquad \leq c(\varepsilon)\|u\|_{H_p(\Omega)}\|v\|_{L^q(\Omega)}.$$

The continuation of f on $H_p(\Omega)\times L^q(\Omega)$ is also denoted by f; for this continuation the same estimates hold.

The Three Lines Theorem of Hadamard gives

$$|\langle B_p^{-\sigma}A_p^{\sigma}u, v\rangle| \leq c\|u\|_{H_p(\Omega)}\|v\|_{L^q(\Omega)},$$

$$\|B_p^{-\sigma}A_p^{\sigma}u\|_{L^p(\Omega)} \leq c\|u\|_{H_p(\Omega)}, \quad u\in D(A_p),$$

$$\|B_p^{-\sigma}u\|_{L^p(\Omega)} \leq c\|A^{-\sigma}u\|_{H_p(\Omega)}, \quad 0\leq\sigma\leq\frac{1}{2p}-\varepsilon, \quad u\in H_p(\Omega).$$

We have made use of the fact that $(B_q^{\sigma})^* = B_p^{\sigma}$ ([Fu]). Of course, if $p=q=2$ we need not use the papers [Gi],[Fu], since then we get from the selfadjointness of A_2, B_2 that even $\|A^{i\tau}\|=1$, $\|B_2^{i\tau}\|=1$. □

The direction $\|A_p^{-\sigma}u\|_{H_p(\Omega)} \leq c\|B_p^{-\sigma}u\|$ is the easy one as will be clear from the proof of theorem III.3.6 later on and may be left to the reader.

§ 3. Local Strong Solvability of the Navier-Stokes Equations

As in the preceding paragraphs n is 2 or 3, Ω is a bounded open set of $\mathbb{R}^n$ whose boundary is of class C^∞. Many of our results remain valid if $n \geq 4$. We will discuss this matter in § 4. For any $u \in (L^1_{loc}(\Omega))^n$ with $\nabla u \in (L^1_{loc}(\Omega))^{n^2}$ and any $v \in (L^1_{loc}(\Omega))^n$ we define $v \cdot \nabla u(x)$ as the vector with components

$$\sum_{\lambda=1}^{n} v_\lambda(x) \frac{\partial u_i}{\partial x_\lambda}(x), \quad 1 \leq i \leq n.$$

$v \cdot \nabla u$ is simply the corresponding mapping from Ω into $\mathbb{R}^n$ but in what follows $v \cdot \nabla u$ is also considered as a mapping between suitable Banach spaces; then v,u vary in certain Banach spaces of integrable functions such that $v \cdot \nabla u$ may be defined as above and we consider the resulting function $v \cdot \nabla u$ as an element of another Banach space.

In its classical formulation the equations of Navier-Stokes

$$\text{(III.3.1)} \quad \frac{\partial u}{\partial t} - \nu\Delta u + u \cdot \nabla u + \nabla \pi = f, \qquad \nabla \cdot u = 0$$

are supposed to determine the velocity $u(t,x)$ and the pressure $\pi(t,x)$ of a viscous incompressible fluid under the influence of an external force f. The fluid fills out the space domain Ω[1] and we suppose that the initial velocity

$$u(0,x) = \varphi(x)$$

and the boundary values of u are prescribed. We assume throughout that the viscosity $\nu > 0$ is constant and that

$$u(t,x) = 0, \quad t > 0, \quad x \in \partial\Omega.$$

---

[1] For the occurrence of the dimension 2 see the comments to this paragraph.

We will see later on that (III.3.1) can be solved locally in t in such a way that u is uniquely determined and more or less regular, but we do not know if these solutions exist for all t. On the other hand, in a weak sense (III.3.1) can be solved for all t but we do not know if the solution is uniquely determined, although a certain sort of local uniqueness holds true (this will be made more precise later on).

Here we start with the following concept: If we formally apply the projection P_p for some $p>1$ to (III.3.1) we get

$$\text{(III.3.2)}\quad u' + A_p u + P_p(u\cdot\nabla u) = P_p f, \qquad u(0) = \varphi$$

as an ordinary differential equation in the Banach space $B = H_p(\Omega)$ if we assume that $u(t) = P_p u(t)$. $-A_p$ is the generator of an analytic semigroup. Thus we can try to apply our former theory to (III.3.2). The pressure π is then determined up to a function of t by (cf. 0.2, p. XVIII)

$$\text{(III.3.3)}\quad \nabla\pi = (I-P_p)\Delta u - (I-P_p)u\cdot\nabla u + (I-P_p)f.$$

We start with an application of the theory developed in II.1. We set

$$M(u) = P_p(u\cdot\nabla u).$$

The norm of $B = H_p(\Omega)$ is denoted by $\|\cdot\|$. The dependence of B on p will not cause any confusion. Observe that $H_p(\Omega)$ is reflexive as was pointed out in the auxiliary propositions 0.2.

<u>Theorem III.3.1:</u> <u>Let</u> $p > \frac{n}{3}$ (> 1 <u>if</u> $n = 2$), $p \neq \frac{3}{2}$. <u>Then the mapping</u>

$$\text{(III.3.4)}\quad M: D(A_p) \to B = H_p(\Omega)$$

<u>defined by</u> $u \to P_p(u\cdot\nabla u)$ <u>fulfills the Lipschitz condition</u> II.1.2, <u>namely</u>

$$\|M(u)-M(v)\| \leq k(C)\|A^{1-\rho}(u-v)\|,$$

$$\|M(u)\| \leq k(C),$$

$$u,v \in D(A_p),\ \|A_p u\|+\|A_p v\| \leq C,$$

<u>for some</u> ρ, $0<\rho<1$. <u>Let</u> $P_p f \in \bigcap_{0<T<+\infty} C^{0,1}([0,T],B)$. <u>Consequently for a</u> $\varphi \in D(A_p)$ <u>there exists a quantity</u> $T(\varphi)$, $0<T(\varphi)\leq +\infty$, <u>with the following properties</u>: <u>There exists a unique</u>

$$u \in \bigcap_{0<T<T(\varphi)} C^1([0,T],B)$$

<u>with</u> $u(t) \in D(A_p)$, $u'(t) \in D(A^{1-\rho})$, $0<t<T(\varphi)$

$$Au \in \bigcap_{0<T<T(\varphi)} C^0([0,T],B),$$

$$A^{1-\rho}u' \in \bigcap_{0<\varepsilon<T<T(\varphi)} C^0([\varepsilon,T],B),$$

$$.^{1-\rho}A^{1-\rho}u'(.) \in \bigcap_{0<T<T(\varphi)} L^\infty((0,T),B),$$

$$u'+A_p u+M(u) = Pf,$$

$$u(0) = \varphi,$$

$$\lim_{t\uparrow T(\varphi)} \|A_p u(t)\| = +\infty,$$

<u>if</u> $T(\varphi)<+\infty$.

<u>Proof:</u> We only have to study the mapping property (II.1.2). The rest follows from theorems II.1.1, III.1.3 and the corollary to theorem II.1.2. On using Hölder's inquality we get

$$\|M(u)-M(v)\| \leq c\|(u-v)\cdot\nabla u\|_{(L^p(\Omega))^n} + c\|v\cdot(\nabla u-\nabla v)\|_{(L^p(\Omega))^n},$$

$$\leq c\|u-v\|_{(L^{pq_1}(\Omega))}\|\nabla u\|_{(L^{pq_2}(\Omega))^{n^2}} +$$

$$+ c\|v\|_{(L^{pq_3}(\Omega))}\|\nabla(u-v)\|_{(L^{pq_4}(\Omega))^{n^2}}$$

with

$$\frac{1}{q_1}+\frac{1}{q_2} = 1, \quad \frac{1}{pq_2} = \frac{1}{p}-\frac{1}{n},$$

$$\frac{1}{q_3}+\frac{1}{q_4} = 1, \quad \frac{1}{pq_3} = \frac{1}{p}-\frac{2}{n}.$$

Thus $\frac{1}{pq_1} = \frac{1}{p}-\frac{1}{pq_2} = \frac{1}{n}$, $\frac{1}{pq_4} = \frac{2}{n}$. Since $p > \frac{n}{3}$ (> 1 if $n = 2$) we have

$$\frac{1}{pq_1} = \frac{1}{n} > \frac{1}{p}-\frac{2}{n}, \quad \frac{1}{pq_4} = \frac{2}{n} > \frac{1}{p}-\frac{1}{n}.$$

Theorem III.2.4 thus shows that

$$\|u-v\|_{(L^{pq_1}(\Omega))^n} \leq c\|A^{1-\rho_1}(u-v)\|,$$

$$\|\nabla u\|_{(L^{pq_2}(\Omega))^{n^2}} \leq c\|Au\|,$$

$$\|v\|_{(L^{pq_3}(\Omega))^{n^2}} \leq c\|Av\|,$$

$$\|\nabla(u-v)\|_{(L^{pq_4}(\Omega))^{n^2}} \leq c\|A^{1-\rho_2}(u-v)\|$$

for some $\rho_1,\rho_2 \in (0,1)$. Taking $\rho = \min(\rho_1,\rho_2)$ our theorem is proved. □

As it turns out the exponents $p=\frac{5}{4}$ for $n=3$ and $p=\frac{4}{3}$ for $n=2$ play an important rôle. Therefore we deal with the exponents $p, n > p > \frac{n}{3}$, $p \neq \frac{3}{2}$ a little bit in detail. First we observe that by Sobolev in any case

$$H^{2,p}(\Omega) \subset L^{n+\varepsilon}(\Omega)$$

for some $\varepsilon > 0$; the imbedding is continuous (even compact, but we will not use this). Thus the solution already constructed in theorem III.3.1 fulfills

$$u \in C^{0}([0,T],(L^{n+\varepsilon}(\Omega))^{n})$$

for any T, $0 < T < T(\varphi)$. As it is natural we assume that $u(t)$ is real valued. Let $n+\varepsilon \geq q \geq \max(2,p)$, where p is as before. Then we consider in $B = H_q(\Omega)$ the operator

$$A_q(t)v = A_q v + P_q(u(t)\cdot\nabla v)$$

with the time independent domain of definition $D(A_q(t)) = D(A_q(0)) = D(A_q)$. We have

$$\begin{aligned}\|A_q(t)v\|_{H_q(\Omega)} &\geq \|A_q v\|_{H_q(\Omega)} - \|P_q(u(t)\cdot\nabla v)\|_{H_q(\Omega)},\\ &\geq \|A_q v\|_{H_q(\Omega)} - c\|u(t)\cdot\nabla v\|_{(L^q(\Omega))^n},\\ &\geq c\|v\|_{2,q} - c'\|u(t)\cdot\nabla v\|_{(L^q(\Omega))^n}.\end{aligned}$$

If $q \leq n$, then we get

$$\text{(III.3.5)}\quad \begin{aligned}\|u(t)\cdot\nabla v\|_{(L^q(\Omega))^n} &\leq c\|u(t)\|_{(L^{n+\varepsilon}(\Omega))^n}\|\nabla v\|_{(L^{\frac{(n+\varepsilon)q}{n+\varepsilon-q}}(\Omega))^{n^2}},\\ &\leq c\|u(t)\|_{(L^{n+\varepsilon}(\Omega))^n}\|v\|_{(W^{1+s,q}(\Omega))^n},\end{aligned}$$

where we have used Hölder's inequality and Sobolev, $s = \frac{n}{n+\varepsilon} < 1$. Theorem III.2.4 and (I.3.3) show that

$$\|v\|_{(W^{1+s,q}(\Omega))^n} \le c\|A_q^{1-\rho'}v\|,$$
$$\le \delta\|A_q v\|_{H_q(\Omega)} + c(\delta)\|v\|_{H_q(\Omega)},$$
$$\le c\delta\|v\|_{2,q} + c(\delta)\|v\|_{H_q(\Omega)}$$

for some $\rho' \in (0,1)$ and all $\delta > 0$. Thus

$$\|A_q(t)v\| \ge c\|v\|_{2,q} - c'\|v\|_{H_q(\Omega)}, \quad 0 \le t \le T.$$

Therefore $A_q(t)$ is closed. Since $q \ge 2$ it follows that

$$(A_q v, v) = \|\nabla v\|^2_{(L^2(\Omega))^{n^2}} \ge c\|v\|^2_{H_2(\Omega)}.$$

Moreover $(P_q(u(t)\cdot\nabla v), v)$ is well defined and we get

$$\begin{aligned} \mathrm{Re}(P_q(u(t)\cdot\nabla v), v) &= \mathrm{Re}(u(t)\cdot\nabla v, P_q v), \\ &= \mathrm{Re}(u(t)\cdot\nabla v, v), \\ &= \mathrm{Re}\sum_{\lambda=1}^{n}(u_i(t)\frac{\partial v_\lambda}{\partial x_i}, v_\lambda), \\ &= \sum_{\lambda=1}^{n}\frac{1}{2}\int_\Omega u_i(t)\frac{\partial|v_\lambda|^2}{\partial x_i}\,dx, \\ &= 0, \end{aligned}$$

by partial integration and since $u_i(t) \in H_2(\Omega)$, $0 \le t \le T$, $\nabla(v_\lambda\overline{v}_\lambda) = \overline{v}_\lambda\nabla v_\lambda + v_\lambda\nabla\overline{v}_\lambda = \overline{v}_\lambda\nabla v_\lambda + \overline{\overline{v}_\lambda\nabla v_\lambda} = 2\,\mathrm{Re}\,\overline{v}_\lambda\nabla v_\lambda \in (L^2(\Omega))^n$; the foregoing calculations are justified by the chain rule (cf. 0.2, auxiliary propositions). Thus we arrive at

$$\text{(III.3.6)}\quad \operatorname{Re}(A_q(t)v,v) \geq c\|v\|^2_{H_2(\Omega)},$$

$$\|A_q(t)v\|_{H_q(\Omega)} \geq c\|v\|_{H_2(\Omega)}.$$

Since $\|v\|_{H_q(\Omega)} \leq \delta\|v\|_{2,q} + c(\delta)\|v\|_{H_2(\Omega)}$, $\delta > 0$, according to the Sobolev inequalities in 0.2, we see that

$$\text{(III.3.7)}\quad \frac{1}{c}\|v\|_{2,q} \leq \|A_q(t)v\|_{H_q(\Omega)} \leq c\|v\|_{2,q},\quad 0 \leq t \leq T.$$

For $\lambda \in \mathbb{C}$, $\operatorname{Re}\lambda \geq 0$, we get with theorem III.1.3 and (III.1.10)

$$\|(\lambda+A_q(t))v\|_{H_q(\Omega)} \geq \|(\lambda+A_q)v\|_{H_q(\Omega)} - \|P_q(u(t)\cdot\nabla v)\|_{H_q(\Omega)},$$

$$\geq c((|\lambda|+1)\|v\|_{H_q(\Omega)} + \|v\|_{2,q}) - \|P_q(u(t)\cdot\nabla v)\|_{H_q(\Omega)}.$$

Employing the same estimates as before for $\|P_q(u(t)\cdot\nabla v)\|_{H_q(\Omega)}$ and (III.3.7) we arrive at

$$\text{(III.3.8)}\quad \|(\lambda+A_q(t))v\|_{H_q(\Omega)} \geq c((|\lambda|+1)\|v\|_{H_q(\Omega)} + \|v\|_{2,q}).$$

Of course all our estimates also hold for the operators $\lambda+A_q^{(\tau)}(t)v = \lambda+A_qv + \tau P_q(u(t)\cdot\nabla v)$, $0 \leq \tau \leq 1$. Assigning to each $w \in (W^{1+s,q}(\Omega))^n$ the uniquely determined solution $\mathcal{T}w$ of

$$(\lambda+A_q)\mathcal{T}w = F - P_q(u(t)\cdot\nabla w) \in H_q(\Omega)$$

we see with theorem III.2.3 that $\mathcal{T}w$ is a compact mapping from $(W^{1+s,q}(\Omega))^n$ into $(H^{2,q}(\Omega))^n$; here F is an arbitrary element from $H_q(\Omega)$. The fixed points of $\tau\mathcal{T}w$, $0 \leq \tau \leq 1$, are uniformly bounded. Thus Schaefer's fixed point theorem (see 0.2, auxiliary propositions) shows that the equation has a solution $v \in D(A_q(0))$, which, according to (III.3.8), is uniquely determined. This also means that all λ with $\operatorname{Re}\lambda \geq 0$ are in the resolvent set of $-A_q(t)$,

$0 \leq t \leq T$, and that the inequality

$$\|(\lambda+A_q(t))^{-1}\| \leq \frac{M'}{|\lambda|+1}$$

holds. Choosing $\varepsilon > 0$ sufficiently small we see from the Nirenberg-Gagliardo inequalities in our auxiliary propositions that

$$\|v\|_{(L^{n+\varepsilon}(\Omega))^n} \leq c\|v\|_{2,p}^{a}\|v\|_{(L^p(\Omega))^n}^{1-a}$$

for some $a \in (0,1)$; e.g. we can take an ε such that $\frac{1}{n+\varepsilon} > -\frac{2}{n}+\frac{1}{p}$ and consequently $\frac{1}{n+\varepsilon} = -\frac{2a}{n}+\frac{1}{p}$ with the quantity $a \in (0,1)$ from the Nirenberg-Gagliardo inequalities. For the solution u of theorem III.3.1 this means

$$\|u(t)-u(s)\|_{(L^{n+\varepsilon}(\Omega))^n} \leq$$

$$\leq c\|u(t)-u(s)\|_{2,p}^{a}\cdot\left\|\frac{u(t)-u(s)}{t-s}\right\|_{(L^p(\Omega))^n}^{1-a}|t-s|^{1-a},$$

$$\leq c(T)|t-s|^{1-a}, \quad 0 \leq t,s \leq T.$$

(III.3.5) and (III.3.7) now show that

$$\text{(III.3.9)} \quad \|(A_q(t)-A_q(r))A_q^{-1}(s)\| \leq c(T)|t-s|^{1-a}, \quad 0 \leq t,s,r \leq T.$$

Thus the $A_q(t)$, $0 \leq t \leq T$, generate an evolution operator $U_q(t,s)$ in the sense of I.2, at least if $\max(2,p) \leq q \leq n$. However, the case $n < q \leq n+\varepsilon$ is even easier to handle with. The reason is that if $q > n$ then

$$\text{(III.3.10)} \quad \|\nabla v\|_{(C^\alpha(\bar{\Omega}))^{n^2}} \leq c\|A_q^{1-\rho'}v\|_{H_q(\Omega)}$$

for some $\alpha \in (0,1)$ as was proved in theorem III.2. 4 . This can be used in estimating $\|P_q(u(t)\cdot\nabla v)\|_{H_q(\Omega)}$. The remaining calculations can be carried through as before, because the imbedding $H^{2,q}(\Omega) \subset C^{1+\alpha}(\overline{\Omega})$ is compact (see lemma 0.2.1). Thus we get in any case the evolution operator $U_q(t,s)$. As for $p \leq q < 2$ we first observe that the inequality (III.3.5) also holds for $q = p$. Thus without using (III.3.6) we get that the $A_p(t) + \Lambda_p$ with

$$A_p(t)g = A_p g + P_p(u(t)\cdot\nabla g),$$
$$g \in D(A_p(t)) = D(A_p(0)) = D(A_p)$$

generate an evolution operator $\tilde{U}_p(t,s)$, where Λ_p is sufficiently large and positive. Using the evolution operator we can derive some additional properties of the solutions constructed in theorem III.3.1. We collect these properties in

Theorem III.3.2: Let f,φ be real valued. Then $u(t)$ is real valued, $0 \leq t < T(\varphi)$. Moreover, if $f \in \bigcap_{0<T<T(\varphi)} C^{\beta}([0,T],(L^{n+\varepsilon}(\Omega))^n)$ for some $\beta,\varepsilon > 0$, then

$$\text{(III.3.11)}\quad \nabla u(t) \in (C^{\alpha}(\overline{\Omega}))^{n^2},$$
$$\|u(t)\|_{(C^{1+\alpha}(\overline{\Omega}))^n} \leq \frac{c(T)}{t^{1-\rho'}},\quad 0 < t \leq T < T(\varphi),$$

for some $\alpha,\rho' \in (0,1)$. As for the higher regularity of u we have the following result: Let moreover $f \in \bigcap_{0<\eta'<T<T(\varphi)} C^0([0,T], (H^{2,n+\varepsilon}(\Omega))^n) \cap C^{0,1}([\eta',T],(L^{n+\varepsilon}(\Omega))^n)$ for some $\varepsilon > 0$. Then

$$u \in \bigcap_{0<\eta'<T<T(\varphi)} C^0([\eta',T],(C^{3+\alpha'}(\overline{\Omega}))^n)$$

and

$$\|u(t)\|_{(C^{3+\alpha'}(\Omega))^n} \leq \frac{c(\varphi,\eta,T)}{(t-\eta)^{1-\rho''}},\quad \eta < t \leq T < T(\varphi),$$

for some $\alpha',\rho'' \in (0,1)$.

Remark: Observe that f is not supposed to fulfill any boundary conditions.

<u>Proof:</u> The proof of the assertion that $u(t)$ is real valued will be postponed to the end of the proof. We choose $q = n+\varepsilon$ throughout this proof. Let us consider the linear equation

$$v' + A_q(t)v = f,$$
$$v(0) = \varphi \in D(A_p) \subset H_q(\Omega).$$

According to theorem I.2.1 this equation has a unique solution $v \in C^0([0,T],H_q(\Omega)) \cap C^1((0,T],H_q(\Omega))$ with $v(t) \in D(A_q(0))$, $0 < t \leq T$, $A_q(.)v(.) \in C^0((0,T],H_q(\Omega))$. The solution v is given by

$$\text{(III.3.12)}\quad v(t) = U_q(t,\eta)\varphi + \int_\eta^t U_q(t,s)f(s)\,ds,\quad 0 \leq \eta \leq t \leq T < T(\varphi).$$

For the difference u-v the following equation in $B = H_p(\Omega)$ holds

$$(u-v)' + A_p(u-v) + P_p(u\cdot\nabla(u-v)) = 0,$$
$$(u-v)(0) = 0.$$

We consider this as an equation for $w = u-v$. Then this equation is rewritten in the form

$$w' + (A_p(t)+\Lambda_p)w - \Lambda_p w = 0,$$

where Λ_p is the constant introduced above in connection with the evolution operator $\tilde{U}_p(t,s)$. On setting $\tilde{w} = e^{-\Lambda_p t}$ we get

$$\tilde{w}' + (A_p(t)+\Lambda_p)\tilde{w} = 0,$$
$$\tilde{w}(0) = 0.$$

From theorem I.2.1 it immediately follows that $\tilde{w}(t) = 0$, $0 \leq t < T(\varphi)$, and therefore

$$u(t) = v(t),\quad 0 \leq t < T(\varphi).$$

The same reasoning as in the proof of theorem III.2.4 shows that

$$\|v\|_{(C^{1+\alpha}(\overline{\Omega}))^n} \leq c\|A_q^{1-\rho'}(t)v\|,\quad 0 \leq t \leq T,$$

$v \in D(A_q^{1-\rho}(t))$ for some $\alpha,\rho' \in (0,1)$ being independent from $t \in [0,T]$. Also the constant c does not depend from $t \in [0,T]$. Equation (III.3.12) then shows that $v(t) = u(t) \in D(A_q^{1-\rho'}(t))$ and that u has in fact the desired property (III.3.11).

As it was stated in 0.2, auxiliary propositions, P_q maps $(C^{l+\beta'}(\bar{\Omega}))^n$ into $H_q(\Omega) \cap (C^{l+\beta'}(\bar{\Omega}))^n$, $l \in \mathbb{N} \cup \{0\}$, $\beta' \in (0,1)$, and the estimate

$$\|P_q u\|_{(C^{l+\beta'}(\bar{\Omega}))^n} \leq c\|u\|_{(C^{l+\beta'}(\bar{\Omega}))^n}$$

holds. This will be used in the sequel. We use again the equation (III.3.12). According to (I.3.7) we have

$$U_q(t,s)x \in D(A_q^{1+\gamma}(t)), \; \gamma < 1-a, \; x \in B, \; t > s,$$

$$A_q^{1+\gamma}(t)U_q(t,0)\varphi \in C^0([\varepsilon',T],B), \; 0 < \varepsilon' < T < T(\varphi).$$

As for the integral term in (III.3.12) we get for $0 < \eta \leq t$:

$$\begin{aligned}
\int_\eta^t U_q(t,s)f(s)\,ds &= \int_\eta^t \frac{\partial}{\partial s}(U_q(t,s)A_q^{-1}(s))f(s)\,ds + \\
&\quad + \int_\eta^t U_q(t,s)A_q^{-1}(s)A_q'(s)A_q^{-1}(s)f(s)\,ds, \\
&= [U_q(t,s)A_q^{-1}(s)f(s)]_\eta^t - \\
&\quad - \int_\eta^t U_q(t,s)A_q^{-1}(s)f'(s)\,ds + \\
&\quad + \int_\eta^t U_q(t,s)A_q^{-1}(s)A_q'(s)A_q^{-1}(s)f(s)\,ds, \\
&= A_q^{-1}(t)f(t) - U_q(t,\eta)A_q^{-1}(\eta)f(\eta) - \\
&\quad - \int_\eta^t U_q(t,s)A_q^{-1}(s)f'(s)\,ds + \\
&\quad + \int_\eta^t U_q(t,s)A_q^{-1}(s)A_q'(s)A_q^{-1}(s)f(s)\,ds.
\end{aligned} \tag{III.3.13}$$

This shows that $u'(t) \in D(A_q^{1-\rho''}(t))$ for any $\rho'' \in (0,1)$, $0 < t \le T$,

$$A_q^{1-\rho''}(.)u'(.) \in C^0((0,T], H_q(\Omega)),$$

$$\|A_q^{1-\rho''}(t)u'(t)\|_{H_q(\Omega)} \le \frac{c}{(t-\eta)^{1-\rho''}}, \quad 0 < \eta < t \le T,$$

provided $A_q(.)x$, $x \in D(A_q)$, is strongly continuously differentiable on $(0,T]$ and $a = 0$. Thus, under our assumption on f, for a sufficiently small ρ'' it follows that $\|A_q(t)u(t)\|_{(C^{1+\alpha'}(\bar{\Omega}))^n} \le c/(t-\eta)^{1-\rho''}$ and that

$$A_q(.)u(.) \in \bigcap_{0<\eta<\eta'<T<T(\varphi)} C^0([\eta',T],(C^{1+\alpha'}(\bar{\Omega}))^n),$$

for some $\alpha \in (0,1)$. We already know from the first part of the proof that $u \cdot \nabla u$ and therefore

$$P_q(u \cdot \nabla u) \in \bigcap_{0<\eta<\eta'<T<T(\varphi)} C^0([\eta',T],(C^{\alpha}(\bar{\Omega}))^n).$$

It follows that $A_q u$ is contained in the same intersection; by the C^α-analogon to the L^p-estimates for the linear stationary problem in [L, 3.5] (the latter we have already employed) we get

$$\text{(III.3.14)} \quad u \in \bigcap_{0<\eta<\eta'<T<T(\varphi)} C^0([\eta',T],(C^{2+\alpha}(\bar{\Omega}))^n).$$

By the C^α-analogon we mean the following theorem (Theorem 5 in [L, 3.5]): Let $f \in (C^{l+\beta}(\bar{\Omega}))^n$ for any $l \in \mathbb{N} \cup \{0\}$ and any $\beta \in (0,1)$. Then there is a unique $u \in (C^{2+l+\beta}(\bar{\Omega}))^n$ and a unique (up to constant) $\pi \in C^{1+l+\beta}(\bar{\Omega})$ with

$$\text{(III.3.15)} \quad -\nu\Delta u + \nabla\pi = f, \; \nabla \cdot u = 0,$$
$$u|\partial\Omega = 0.$$

Moreover, the following estimate holds:

$$\text{(III.3.16)} \quad \|u\|_{(C^{2+l+\beta}(\bar{\Omega}))^n} + \|\nabla\pi\|_{(C^{l+\beta}(\bar{\Omega}))^n} \le c\|f\|_{(C^{l+\beta}(\bar{\Omega}))^n}.$$

We have not mentioned this theorem separately because it is only used at this point. For $l=0$ we in fact arrive at (III.3.14). $u\cdot\nabla u$ and therefore

$$P_q(u\cdot\nabla u)\in \bigcap_{0<\eta<\eta'<T<T(\varphi)} C^0([\eta',T],(C^{1+\alpha}(\bar{\Omega}))^n),$$

and the same device as before (but now for $l=1$) gives

$$u\in \bigcap_{0<\eta<\eta'<T<T(\varphi)} C^0([\eta',T],(C^{3+\alpha}(\bar{\Omega}))^n).$$

The differentiability of $A_q(.)x$ simply follows (III.3.12), since $u=v\in C^1((0,T],H_q(\Omega))$, $0<T<T(\varphi)$, and therefore also $a=0$.

It remains to prove that $u(t)$ is real valued if φ,f are so. Since the iteration procedure in the proof of theorem II.1.1 is based on the consideration of the mapping ($A=A_p$)

$$\mathcal{T}w(t) = e^{-tA}\varphi - \int_0^t e^{-(t-s)A}(M(w(s))-f(s))\,ds$$

we only have to show that M is real valued if w is so and that $e^{-tA}x$ is real valued if $x\in B=H_p(\Omega)$ is real valued; the second term in the integral equation is caused by the fact that we have an external force f in the Navier-Stokes equations. $e^{-tA}x$ is the solution of $u'+Au=0$, $u(0)=x$, in the sense of theorem I.2.1. Since P_p maps the real valued functions of $(L^p(\Omega))^n$ into itself we get the first assertion and

$$v_1'+Av_1 = 0,\ v_1(0)=x$$
$$v_2'+Av_2 = 0,\ v_2(0)=0$$

with $v_1=\operatorname{Re} u$, $v_2=\operatorname{Im} u$. The equations for v_1,v_2 have to be taken in the sense of theorem I.2.1, and therefore $v_2=0$. Our theorem is proved. □

It may be noted that the assumptions on f could be weakened, in particular as it concerns the regularity conditions on f in the last part of the preceding theorem.

So far we have considered initial values φ being in the domain of definition of A_p. Now we turn to the question whether local strong solutions exist if φ is only in B. The basic ideas for the answer were given in an abstract way in II.3; as a matter of fact, for the Navier-Stokes equations it is more useful to treat these questions for $n \geq 4$ than for $n = 2,3$. Nevertheless we remain at the cases $n = 2,3$ and we will describe in § 4 the necessary modifications for $n \geq 4$; as it turns out, these are in no way severe. Since we also want to treat the question if u is in the domain of certain powers of A greater than 1 we want to start with the analyticity of the mapping $P_p(u\cdot\nabla u)$ in the sense of II.2.

<u>Theorem III.3.3:</u> <u>Let</u> $p > n$, <u>say</u> $p = n+\varepsilon$ <u>for some</u> $\varepsilon > 0$. <u>Let</u> $B = H_p(\Omega)$. <u>Then</u> $M(u) = P_p(u\cdot\nabla u) = P(u\cdot\nabla u)$ <u>has the following properties:</u>

$$M \in A(D(A_p^{1-\rho_1}),B)$$

<u>for every</u> ρ_1, $0 \leq \rho_1 < \frac{1}{2}-\frac{n}{2p}$, <u>and</u>

$$M \in A(D(A_p),D(A_p^{\rho_2}))$$

<u>for every</u> $\hat{\rho}_2$, $0 \leq \hat{\rho}_2 < \frac{1}{2p}$.

<u>Proof:</u> First we remark that, according to 0.2, the projection operator $P = P_p$ maps $(H^{1,p}(\Omega))^n$ into itself and an inequality

$$\|Pu\|_{(H^{1,p}(\Omega))^n} \leq c\|u\|_{(H^{1,p}(\Omega))^n}$$

holds. According to theorem III.2.4 the following estimate holds (Observe that $p > n$):

$$\text{(III.3.17)} \quad \|M(u)\| \leq c(\int_\Omega |u\cdot\nabla u|^p\, dx)^{\frac{1}{p}} \leq c\|\nabla u\|_{(C^0(\bar{\Omega}))^{n^2}}\|u\|_{(L^p(\Omega))^n},$$

$$\leq c\|A^{1-\rho_1}u\|\,\|u\|,\ \rho_1 \in [0,\tfrac{1}{2}-\tfrac{n}{2p});$$

$\|.\|$ is the norm of B. The same argument shows that the Lipschitz condition

$$\text{(III.3.18)} \quad \|M(u)-M(v)\| \leq c[(\|u\|+\|v\|)\|A^{1-\rho_1}(u-v)\| + (\|A^{1-\rho_1}u\|+\|A^{1-\rho_1}v\|)\|u-v\|]$$

holds. As for the analyticity we have for almost all $x \in \Omega$ and all $u, v \in D(A^{1-\rho_1})$

$$(u+zv)(x)\cdot\nabla(u+zv)(x) = \frac{1}{2\pi i}\int_{\partial K} \frac{(u+\zeta v)(x)\cdot\nabla(u+\zeta v)(x)}{\zeta - z}\,d\zeta$$

where K is a suitable ball of radius 1 with center z_o; the reason is simply that for almost all $x \in \Omega$ the vector function $z \mapsto (u+zv)(x)\cdot\nabla(u+zv)(x)$ is an entire function. Because of (III.3.18) the last integral can be understood as a Bochner integral in $B = H_p(\Omega)$; thus we get

$$(u+zv)\cdot\nabla(u+zv) = \frac{1}{2\pi i}\int_{\partial K} \frac{(u+\zeta v)\cdot\nabla(u+\zeta v)}{\zeta - z}\,d\zeta$$

and therefore the assertion: $M \in A(D(A^{1-\rho_1}),B)$. If $u \in D(A_p) = (H^{2,p}(\Omega) \cap \overset{o}{H}{}^{1,p}(\Omega))^n \cap H_p(\Omega)$ then it follows from the chain rule of 0.2 that $u\cdot\nabla u \in (H^{1,p}(\Omega))^n$, and the derivatives are formed by formal differentiation. In doing so we get with the Gagliardo-Nirenberg inequality of our auxiliary propositions

$$\|u\cdot\frac{\partial}{\partial x_i}\nabla u\|_{(L^p(\Omega))^n} \leq c\|u\|_{(H^{2,p}(\Omega))^n}\|u\|_{(C^o(\bar{\Omega}))^n},$$

$$\leq c\|u\|^{1+\frac{n}{2p}}_{(H^{2,p}(\Omega))^n}\|u\|^{1-\frac{n}{2p}}_{(L^p(\Omega))^n},$$

$$\|\frac{\partial}{\partial x_i}u\cdot\nabla u\|_{(L^p(\Omega))^n} \leq c\|u\|_{(H^{2,p}(\Omega))^n}\|u\|_{(C^o(\bar{\Omega}))^n}.$$

Thus we arrive at

$$\text{(III.3.19)}\quad \|M(u)\|_{(H^{1,p}(\Omega))^n} \le c\|u\cdot\nabla u\|_{(H^{1,p}(\Omega))^n}$$

$$\le c\|u\|^{1+\frac{n}{2p}}_{(H^{2,p}(\Omega))^n}\cdot\|u\|^{1-\frac{n}{2p}}_{(L^p(\Omega))^n}$$

$$\le c\|A_p u\|^{1+\frac{n}{2p}}\|u\|^{1-\frac{n}{2p}}.$$

As before it is shown that the mapping: $u\to P(u\cdot\nabla u)$ defines a mapping from the class $A(D(A_p),(H^{1,p}(\Omega))^n\cap B)$, and therefore from the class $A(D(A_p),(H^{\frac{1}{p},p}(\Omega))^n\cap B) = A(D(A_p),(\overset{o}{H}{}^{\frac{1}{p},p}(\Omega))^n\cap B)$. Theorem III.2.4 completes the proof. □

From Theorem III.3.3 we want to draw some consequences:

<u>Proposition III.3.1:</u> <u>For any</u> $\hat\rho_2 \in [0,\frac{1}{2p})$ <u>there exists an</u> $\varepsilon' = \varepsilon'(\hat\rho_2)\in(0,1)$ <u>such that</u>

$$M\in A(D(A_p^{1-\varepsilon'}),D(A_p^{\hat\rho_2})).$$

<u>Proof:</u> We set $A=A_p$. For η, $\eta>0$, and $\hat\rho_2'$, $0<\hat\rho_2'<\frac{1}{2p}$, the nonlinearity M is a mapping from $D(A^{\gamma+\eta})$ into $D(A^{\delta-\eta})$ with

$$\gamma = \gamma(\sigma) = 1-\sigma+\sigma(1-\rho_1),\ \gamma+\eta\le 1,$$

$$\delta = \delta(\sigma) = (1-\sigma)\hat\rho_2'.$$

This follows from theorem II.2.1. We choose a fixed $\hat\rho_2\in(0,\frac{1}{2p})$. Let $\hat\rho_2'$ be an element from $(\hat\rho_2,\frac{1}{2p})$. Then there is a $\sigma_2\in(0,1)$ with

$$\eta+\hat\rho_2 = (1-\sigma_2)\hat\rho_2'$$

if $\eta > 0$ is sufficiently small. Moreover we have

$$1-\sigma_2+\sigma_2(1-\rho_1) < 1.$$

Therefore, for sufficiently small $\eta > 0$, we get that also

$$1-\varepsilon' = 1-\sigma_2+\sigma_2(1-\rho_1)+\eta < 1.$$

According to the proof of theorem II.2.1 we even have

$$M \in A(D(A^{1-\varepsilon'}), D(A^{(1-\sigma_2)\hat{\rho}_2'-\eta}))$$

$$= A(D(A^{1-\varepsilon'}), D(A^{\hat{\rho}_2})). \qquad \square$$

We remark that from (II.3.19) it follows that

$$\text{(III.3.20)} \quad \|A^{\hat{\rho}_2}M(u)\| \leq c(\|A^{1-\varepsilon'}u\|^{1+\rho_4}+\|A^{1-\varepsilon'}u\|)(\|u\|+\|u\|^{1-\rho_4})$$

for some $\rho_4 \in [0,1]$. It is easy to see that the second estimate in (II.3.5) can be replaced by the following one:

$$\text{(III.3.21)} \quad \|A^{\kappa-(1-\rho_2)}M(u)\| \leq c(\|A^{\kappa}u\|^{1+\rho_4}+\|A^{\kappa}u\|)(\|u\|+1)$$

in the sense that this will not affect the validity of theorem II.3.5.

Now we can prove the following theorem on the local solvability of the Navier-Stokes equations:

<u>Theorem III.3.4:</u> <u>Let</u> $p > n$. <u>Let</u> $\hat{\rho}_2 \in (0,\frac{1}{2p})$. <u>Let</u> $f \in C^{\frac{1}{2p}}([0,T], (L^p(\Omega))^n) \cap C^0((0,T],(H^{\frac{1}{p},p}(\Omega))^n)$, $T > 0$. <u>Let</u> $\varphi \in B = H_p(\Omega)$. <u>There exists a</u> $T(\varphi)$, $+\infty \geq T(\varphi) > \infty$, <u>such that the integral equation</u>

$$u(t) = e^{-tA}\varphi - \int_0^t e^{-(t-s)A}(M(u(s))-f(s))\, ds$$

<u>has a unique solution</u>

$$u \in C^{o}([0,T(\varphi)),B)$$

<u>with</u>

$$u(t) \in D(A^{1+\hat{\rho}_2}),\ 0 < t < T(\varphi),$$

$$A^{1+\hat{\rho}_2}u(.) \in C^{o}((0,T(\varphi)),B),$$

$$u \in C^{1}((0,T(\varphi)),B),$$

$$u'(t) \in D(A^{\hat{\rho}_2}),\ 0 < t < T(\varphi),$$

$$A^{\hat{\rho}_2}u'(.) \in C^{o}((0,T(\varphi)),B),$$

$$.^{1-\rho_1}A^{1-\rho_1}u(.) \in \bigcap_{0<T<T(\varphi)} L^{\infty}((0,T),B),$$

$$u' + Au + M(u) = Pf,$$

$$u(0) = \varphi,$$

<u>where</u> $A = A_p$, $P = P_p$; <u>if</u> $T(\varphi) < +\infty$, <u>then</u>

$$\lim_{t\uparrow T(\varphi)} \|u(t)\| = +\infty.$$

<u>Proof:</u> First of all the quantity $T(\varphi)$ is determined according to theorem II.3.3. The rest follows from theorems II.3.3, II.3.5, II.3.9. For that purpose one has to insert in theorem II.3.5

$$\kappa = 1-\varepsilon',$$

$$\hat{\rho}_2 = \kappa-(1-\rho_2), \text{ i.e.}$$

$$\rho_2 = 1+\hat{\rho}_2-\kappa,$$

where ε', $\hat{\rho}_2$ are the quantities from proposition III.3.1. □

We next turn to the proof of the theorem of Fujita-Kato [FK] on the local solvability of the Navier-Stokes equations for $n=3$; this proof is an easy application of theorem II.3.6. The restriction $n=3$ is important now.

Theorem III.3.5: Let $p=2$. Let $A=A_2$, let B be the Hilbert space $H=H_2(\Omega)$. Let $f\in C^{\frac{1}{4}}([0,T],H)\cap C^0((0,T],(H^{\frac{1}{2},2}(\Omega))^3)$, $T>0$. Let $\hat{\rho}_2$ be an element from $(0,\frac{1}{4})$. There exists a $T(\varphi)$, $+\infty\geq T(\varphi)>0$, with the following properties: The integral equation

$$u(t) = e^{-tA}\varphi - \int_0^t e^{-(t-s)A}(M(u(s))-f(s))\,ds$$

has a unique solution

$$u\in C^0([0,T(\varphi)),H)$$

with

$$u(t)\in D(A^{\frac{1}{4}}),\ 0\leq t<T(\varphi),$$

$$A^{\frac{1}{4}}u(.)\in C^0([0,T(\varphi)),H).$$

Moreover

$$u(t)\in D(A^{1+\hat{\rho}_2}),\ 0<t<T(\varphi)$$

$$A^{1+\hat{\rho}_2}u(.)\in C^0((0,T(\varphi)),H),$$

$$u\in C^1((0,T(\varphi)),H),$$

$$u'(t)\in D(A^{\hat{\rho}_2}),\ 0<t<T(\varphi),$$

$$A^{\hat{\rho}_2}u'(.)\in C^0((0,T(\varphi)),H),$$

$$\cdot^{\frac{1}{2}} A^{\frac{3}{4}} u(\cdot) \in \bigcap_{0<T<T(\varphi)} L^{\infty}((0,T),H),$$

$$u' + Au + M(u) = Pf,$$

$$u(0) = \varphi.$$

_If_ $T(\varphi) < +\infty$, _then_

$$\lim_{t\uparrow T(\varphi)} \|A^{\frac{3}{4}} u(t)\| = +\infty.$$

$\|\cdot\|$ _is the norm of_ $(L^2(\Omega))^n$.

Proof: We have already shown in theorem III.1.2 that A is selfadjoint and > 0. Considering simultaneously the selfadjoint operator $B = -\nu\Delta$ in $(L^2(\Omega))^3$ with domain of definition $(H^{2,2}(\Omega))^3 \cap (\overset{\circ}{H}{}^{1,2}(\Omega))^3$ we see that

$$(Au,u) = \|A^{\frac{1}{2}}u\|^2 = \|B^{\frac{1}{2}}u\|^2 = (Bu,u) = \|\nabla u\|^2,$$

$u \in D(A)$. As it follows from theorem III.2. 5 the domain of definition $D(A^{\frac{1}{2}})$ of $A^{\frac{1}{2}}$ is exactly the space $\{u \mid u \in (\overset{\circ}{H}{}^{1,2}(\Omega))^3,\ \nabla\cdot u = 0\}$. Taking for T the imbedding $H = H_2(\Omega) \ni u \to u \in (L^2(\Omega))^3$ we have seen from the Heinz-Kato inequality in our theorem III.2.5 that

$$\|B^{\rho}u\| \le \|A^{\rho}u\|,\ u \in D(A^{\frac{1}{2}}),\ 0 \le \rho \le \tfrac{1}{2}.$$

As we have also stated there that

$$D(B^{\rho}) \subset (H^{2\rho,2}(\Omega))^3,\ 0 \le \rho \le \tfrac{1}{4},$$

with a continuous imbedding, where, of course, $D(B^{\rho})$ is equipped with the graph norm of B^{ρ}; in fact the norms are equivalent with the exception of $\rho = \frac{1}{4}$. Using Hölders inequality we get then with the aid of (I.3.3)

$$\begin{aligned}\|M(u)\| &\le c\|u\|_{(L^6(\Omega))^3}\|\nabla u\|_{(L^3(\Omega))^3},\\ &\le c\|u\|_{(H^{1,2}(\Omega))^3}\|A^{-\frac{1}{2}}A^{\frac{1}{2}}u\|_{(H^{3/2,2}(\Omega))^3},\\ &\le c\|A^{\frac{1}{2}}u\|(\|A^{\frac{1}{2}}u\|_{(H^{1/2,2}(\Omega))^3}+\|u\|_{(H^{1/2,2}(\Omega))^3}),\\ &\le c\|A^{\frac{1}{2}}u\|\,\|A^{\frac{3}{4}}u\| \le c\|A^{\frac{3}{4}}u\|^{\frac{3}{2}}\|A^{\frac{1}{4}}u\|^{\frac{1}{2}},\end{aligned}$$

$u \in D(A^{\frac{3}{4}})$. For $u,v \in D(A^{\frac{3}{4}})$ a corresponding Lipschitz estimate

$$\begin{aligned}\|M(u)-M(v)\| \le\; & c(\|A^{\frac{1}{4}}u\|+\|A^{\frac{1}{4}}v\|)^{\frac{1}{2}}(\|A^{\frac{3}{4}}u\|+\|A^{\frac{3}{4}}v\|)^{\frac{1}{2}}\|A^{\frac{3}{4}}(u-v)\| +\\ &+ c(\|A^{\frac{3}{4}}u\|+\|A^{\frac{3}{4}}v\|)\|A^{\frac{3}{4}}(u-v)\|^{\frac{1}{2}}\|A^{\frac{1}{4}}(u-v)\|^{\frac{1}{2}}\end{aligned}$$

is easily derived by using the same method. Thus we see that the assumptions of theorem II.3.6 are fulfilled if we insert

$$\delta = \frac{1}{4},$$

$$\rho_o = \frac{1}{2},$$

$$\rho_1 = \frac{1}{4};$$

in particular then the relation $1-\rho_1-\frac{\rho_1}{\rho_o} = \frac{1}{4} = \delta < 1-\rho_1$ holds. Thus theorem II.3.6 furnishes $T(\varphi)$ and the first part of our theorem. Next we deal with the higher regularity. We want to apply theorem II.3.7. We can apply the chain rule to $\frac{\partial}{\partial x_i}(u\cdot\nabla u)$ since $H^{2,2}(\Omega) \subset C^o(\bar{\Omega})$ if $n=3$, $H^{1,2}(\Omega) \subset L^6(\Omega)$ if $n=3$ with continuous imbeddings, and consequently

$$(\int_\Omega |u|^2|\nabla^2 u|^2\,dx)^{\frac{1}{2}} \le c\|Au\|^2,$$

$$(\int_\Omega |\nabla u|^2|\nabla u|^2\,dx)^{\frac{1}{2}} \le c\|Au\|^2.$$

Thus $M(u)$ is in $(H^{1,2}(\Omega))^3$ if $u \in D(A)$ ($u\cdot\nabla u$ is even contained in $(\overset{\circ}{H}{}^{1,2}(\Omega))^3$). The chain rule also shows that

$$M \in A(D(A),(H^{1,2}(\Omega))^3).$$

Thus we get in particular that

$$M \in A(D(A),D(A^{\hat{\rho}_2'}))$$

for any $\hat{\rho}_2' \in (0,\frac{1}{4})$ (cf. theorem III.2. 5). Let $\hat{\rho}_2$ be arbitrary from $(0,\frac{1}{4})$ but fixed, let $\hat{\rho}_2 < \hat{\rho}_2' < \frac{1}{4}$. We have

$$\|A^{\kappa-(1-\rho_2)} M(u)\| \le c\|A^{\kappa}u\|^{1+\rho_4}(\|u\|+1)$$

with $\kappa = 1$, $\kappa-(1-\rho_2) = \rho_2'$, i.e. $\rho_2 = \hat{\rho}_2'$, $\rho_4 = 1$. From the first part of the proof we already know that

$$\|M(u)\| \le c\|A^{\frac{1}{2}}u\|\,\|A^{\frac{3}{4}}u\|;$$

(I.3.3) shows that

$$\|M(u)\| \le c\|A^{\frac{3}{4}}u\|^{\frac{5}{3}}(\|u\|+1).$$

We set $\rho_1 = \frac{1}{4}$, $\rho_o = \frac{2}{3}$. From theorem II.3.6 it also follows that

$$A^{\frac{3}{4}}u(.) \in \bigcap_{0<\varepsilon<T<T(\varphi)} C^o([\varepsilon,T],H),$$

$$\|A^{\frac{3}{4}}u(t)\| \le \frac{c(T)}{t^{1/2}},\quad 0 < t < T;$$

thus $A^{\frac{3}{4}}u(.) \in \bigcap_{0<T<T(\varphi)} L^{1+\rho_o}((0,T),B) = \bigcap_{0<T<T(\varphi)} L^{5/3}((0,T),B)$.

Thus the assumptions of theorem II.3.7 are fulfilled with κ,ρ_2, ρ_1,ρ_o as above ; but the theorems II.3.7, II.3.6 immediately give the remaining part of theorem III.3.5. □

We want to make three remarks on the preceding theorems. As it was pointed out in 0.2, the assumption: $\varphi \in D(A^{1/4})$ means that a boundary condition is imposed on φ whereas this is not needed in theorem III.3.4. On the other hand we have

$$D(A^{\frac{1}{4}}) \subset (L^3(\Omega))^3 = (L^n(\Omega))^n$$

with a continuous imbedding and 3 is the marginal exponent. The theorem to follow may be considered as an improvement of both of the theorems III.3.4 and III.3.5. The second remark we make is that of course the solution constructed in theorem III.3.5 can be identified for t, $0 < t < T(\varphi)$, with the solution constructed in III.3.4 if the assumptions on f are adjusted in a proper way. The third remark may be the most interesting one; it concerns the behaviour of u if one approaches $T(\varphi)$ in theorem III.3.5, provided $T(\varphi) < +\infty$. The corresponding assertion in theorem III.3.5 can be strengthened considerably. For that purpose we have to go back to the proof of theorem II.3.6. The length $\tilde{T}$ of the interval on which we have constructed the solution in the proof of theorem II.3.6 depends on

$$\sup_{0 \le t \le \tilde{T}} \| e^{-tA} A^{\delta} \varphi \|$$

and the rate of convergence to 0 of

$$\sup_{0 < t \le \tilde{T}} \| t^{1-\rho_1-\delta} A^{1-\rho_1} e^{-tA} \varphi \|$$

if $t \to 0$. Thus it is a contradiction to assume that $T(\varphi) < +\infty$ and simulatneously that

$$\text{(III.3.22)} \quad \sup_{0 \le t < T(\varphi)} \| A^{\delta} u(t) \| \le D < +\infty$$

and

$$\text{(III.3.23)} \quad w_{\delta} = \sup_{\substack{0 \le t \le \delta \\ 0 < T < T(\varphi)}} \| t^{1-\rho_1-\delta} A^{1-\rho_1} e^{-tA} u(T) \| \to 0$$

for $\delta \to 0$. Thus we get that $T(\varphi)$ is $< +\infty$ if

(III.3.24) $\lim\limits_{t \uparrow T(\varphi)} \|A^{\delta} u(t)\| = +\infty$

or

(III.3.25) $\left\{ \begin{array}{l} w_{\delta} = \sup\limits_{\substack{0<t<\delta \\ 0<T<T(\varphi)}} \|t^{1-\rho_1-\delta} A^{1-\rho_1} e^{-tA} u(T)\| \\ \\ \text{is } \underline{\text{not}} \text{ convergent to } 0 \text{ if } \delta \to 0. \end{array} \right.$

The conditions (III.3.24) and (III.3.25) can be unified in the following way: If $A^{\delta}u(.)$ is uniformly continuous in t on $[0,T(\varphi))$ then $A^{\delta}u(.)$ can be continued in a unique way on $[0,T(\varphi)]$ such that $A^{\delta}u(.)$ is continuous on $[0,T(\varphi)]$. Moreover there is a sequence $u_k \in C^0([0,T],D(A^{1+\delta}))$ with

$$u_k \to u \text{ in } C^0([0,T],D(A^{\delta}))$$

and we have

$$\begin{aligned} &\|t^{1-\rho_1-\delta} A^{1-\rho_1} e^{-tA} u(T)\| \\ &\leq \|t^{1-\rho_1-\delta} A^{1-\rho_1-\delta} e^{-tA} A^{\delta}(u(T)-u_k(T))\| + \\ &\quad + \|t^{1-\rho_1-\delta} A^{-\rho_1-\delta} e^{-tA} A^{1+\delta} u_k(T)\|, \end{aligned}$$

from which it follows that (III.3.23) holds. Thus we get that $T(\varphi) < +\infty$ if

(III.3.26) $A^{\delta}u$ is not uniformly continuous on $[0,T(\varphi))$.

Now we turn to the theorem already announced.

Theorem III.3.6: Let $\varphi \in H_n(\Omega)$. *Let* $f \in C^{\frac{1}{2p}}([0,T],L^p(\Omega)) \cap C^0((0,T],H^{\frac{1}{p},p}(\Omega))$ *for all* $T>0$ *and some* $p>n$. *Then there is a* $T(\varphi) \in (0,+\infty]$ *with the following properties: There is one and only one*

$$u \in \bigcap_{0<T<T(\varphi)} C^0([0,T],H_n(\Omega))$$

with

$$u(0)=\varphi,\ u(t) \in D(A_n^{\frac{1}{2}-\varepsilon}),\ 0<t<T(\varphi),\ 0<\varepsilon<\frac{1}{2},$$

$$\cdot^{\frac{1}{2}-\varepsilon} A_n^{\frac{1}{2}-\varepsilon} u(.) \in \bigcap_{\substack{0<T<T(\varphi),\\ 0<\varepsilon<\frac{1}{2}}} C^0((0,T),H_n(\Omega)) \cap L^\infty((0,T),H_n(\Omega)),$$

$$u(t) = e^{-tA_n}\varphi - \int_0^t e^{-(t-s)A_n}(M(u(s))-f(s))\,ds,$$

$0<t<T(\varphi)$. *The last integral equation has to be taken in* $H_n(\Omega)$. *If* $T(\varphi)<+\infty$, *then*

(III.3.27) u *is not uniformly continuous on* $(0,T(\varphi))$ *with respect to* t *in the* $H_n(\Omega)$*-norm.*

As for the higher regularity of u *the following result holds:*

$$u \in \bigcap_{0<T<T(\varphi)} C^1((0,T),H_p(\Omega)),$$

$$u(t) \in D(A_p^{1+\rho}),\ 0<t<T(\varphi),\ 0<\rho<\frac{1}{2p},$$

$$A_p^{1+\rho}u(.) \in \bigcap_{0<T<T(\varphi)} C^0((0,T),H_p(\Omega)),$$

$$u' + A_p u + M(u) = P_p f$$

on $(0,T(\varphi))$.

Proof: For the proof we need the theorem of Giga on the fractional powers of the Stokes operator; this has been mentioned after our theorem III.2.6.

First we give an estimate for $A_n^{-\frac{1}{2}}M(u)$. Let $\tilde{\varphi} \in (C_o^{\infty}(\Omega))^n$. Then

$$(u\cdot\nabla u,\tilde{\varphi}) = \sum_{l=1}^{n} \Big(\sum_{i=1}^{n} u_i \frac{\partial u_l}{\partial x_i}, \tilde{\varphi}_l \Big)$$

$$= - \sum_{i,l=1}^{n} \Big(u_i u_l, \frac{\partial \tilde{\varphi}_l}{\partial x_i}\Big)$$

$$= - (u\cdot u, \nabla\tilde{\varphi}),$$

where $u\cdot u$ is the vector with components $u_i \cdot u_l$, $1 \leq i,l \leq n$. Here $u \in D(A_n)$ which also means that by Sobolev all expressions are well defined. From this estimate we get

$$|(u\cdot\nabla u,\tilde{\varphi})| \leq c\|u\|^2_{(L^{2n}(\Omega))^n}\|\tilde{\varphi}\|_{(H^{1,n/(n-1)}(\Omega))^n}$$

and

$$\|u\cdot\nabla u\|_{(H^{-1,n}(\Omega))^n} \leq c\|u\|^2_{(L^{2n}(\Omega))^n}.$$

The same calculations show that

$$\|u\cdot\nabla u - v\cdot\nabla v\|_{(H^{-1,n}(\Omega))^n} \leq c(\|u\|_{(L^{2n}(\Omega))^n} + \|v\|_{(L^{2n}(\Omega))^n})\cdot$$

$$\cdot\|u-v\|_{(L^{2n}(\Omega))^n}.$$

Assuming for a moment that $\|A_n^{-\frac{1}{2}}Pf\|_{H_n(\Omega)} \leq c\|f\|_{(H^{-1,n}(\Omega))^n}$, $f \in (L^n(\Omega))^n$, we then get with the aid of Giga's result that

$$(\text{III.3.28})\quad \begin{cases} \|A_n^{-\frac{1}{2}}M(u)\|_{H_n(\Omega)} \le c\|A_n^{\frac{1}{4}}u\|^2_{H_n(\Omega)}, \\ \|A_n^{-\frac{1}{2}}(M(u)-M(v))\|_{H_n(\Omega)} \le c(\|A_n^{\frac{1}{4}}u\|_{H_n(\Omega)} + \|A_n^{\frac{1}{4}}v\|_{H_n(\Omega)})\cdot \\ \qquad \cdot\|A_n^{\frac{1}{4}}(u-v)\|_{H_n(\Omega)}. \end{cases}$$

Theorem III.1.3 shows that

$$(\text{III.3.29})\quad (\lambda+A_p)^{-1}u = (\lambda+A_q)^{-1}u,\ u\in H_p(\Omega),$$

if $p \ge q > 1$; λ is as in theorem III.1.3. Therefore $A_p^{\frac{1}{2}}u = A_q^{\frac{1}{2}}u$, $u \in D(A_p^{\frac{1}{2}}) \supset D(A_q^{\frac{1}{2}})$ by the definition of the fractional powers of A. Thus we get for $\tilde{\varphi} \in (C_o^\infty(\Omega))^n$, $\nabla\cdot\tilde{\varphi} = 0$

$$|(A_n^{-\frac{1}{2}}Pf, A_{n/(n-1)}^{\frac{1}{2}}\tilde{\varphi})| = |(f,\tilde{\varphi})|$$

$$\le \|f\|_{(H^{-1,n}(\Omega))^n}\|\tilde{\varphi}\|_{(H^{1,n/(n-1)}(\Omega))^n}$$

$$\le c\|f\|_{(H^{-1,n}(\Omega))^n}\|A^{\frac{1}{2}}_{\frac{n}{n-1}}\tilde{\varphi}\|,$$

where we have used again Giga's result on the fractional powers of A. This in particular shows that

$$\|A_n^{-\frac{1}{2}}Pf\| \le c\|f\|_{(H^{-1,n}(\Omega))^n}.$$

Thus (III.3.28) is justified. Now we use theorem II.3.8 in $B = H_n(\Omega)$. Setting $\rho_1 = 0$, $\delta = \frac{1}{2}$, $\tilde{M}(u) = M(A_n^{\frac{1}{2}}u)$ we see that the Lipschitz conditions in II.3.8 are fulfilled if we set $\rho_1' = \frac{1}{4}$; this

$$0 < 1-2\rho_1' = \frac{1}{2} = \delta < \frac{3}{4} = 1-\rho_1'.$$

Thus theorem II.3.8 furnishes the first part of theorem III.3.6 with the exception of the characterization of the case: $T(\varphi) < +\infty$. In fact we have solved the integral equation

$$r(t) = A_n^{-\delta}u(t) = e^{-tA_n}A_n^{-\frac{1}{2}}\varphi - \int_0^t e^{-(t-s)A_n}(A_n^{-\frac{1}{2}}(M(A_n^{\frac{1}{2}}A_n^{-\frac{1}{2}}u(s))-P_nf(s))\,ds$$

$$= e^{-tA_n}A_n^{-\frac{1}{2}}\varphi - \int_0^t e^{-(t-s)A_n}(A_n^{-\frac{1}{2}}M(A_n^{\frac{1}{2}}v(s))-P_nf(s))\,ds$$

with the nonlinearity $A_n^{-\frac{1}{2}}M(A_n^{\frac{1}{2}}u)$ which can be defined on $D(A_n^{\frac{3}{4}})$ by (III.3.28); on $D(A_n^{\frac{3}{4}})$ this mapping fulfills the assumptions of theorem II.3.8. Thus we can use (III.3.26) to characterise the case $T(\varphi) < +\infty$. This immediately gives the result desired. As for the higher regularity we want to use theorem III.3.4. Since $u(t) \in D(A_n^{\frac{1}{2}-\varepsilon})$, $0 < t < T(\varphi)$, for any $\varepsilon > 0$ we get by Sobolev that in particular

$$u \in C^0((0,T(\varphi)),H_p(\Omega))$$

for the $p > n$ already introduced in the present theorem. If $\tilde{\eta}$ is arbitrary in $(0,T(\varphi))$, then u fulfills the integral equation

$$\text{(III.3.30)}\quad A_n^{-\frac{1}{2}}u(t) = e^{-tA_n}u(\tilde{\eta}) - \int_{\tilde{\eta}}^t e^{-(t-s)A_n}(A_n^{-\frac{1}{2}}M(u(s))-A_n^{-\frac{1}{2}}P_nf(s))$$

in $H_n(\Omega)$, $\tilde{\eta} \le t < T(\varphi)$. Now we reconstruct u by solving the equation

$$\tilde{u}(t) = e^{-tA_p} u(\tilde{\eta}) - \int_{\tilde{\eta}}^{t} e^{-(t-s)A_p}(M(\tilde{u}(s)) - P_p f(s))\, ds$$

on $[\tilde{\eta},\tilde{T}+\tilde{\eta}]$ for a certain $\tilde{T}>0$ by using theorem III.3.4. From (III.3.29) it follows that $\tilde{u}$ also satisfies (III.3.30) on $[\tilde{\eta},\tilde{T}+\tilde{\eta}]$. Because of the regularity properties of $\tilde{u}$ (stated in theorem III.3.4) we get that

(III.3.31) $u(t) = \tilde{u}(t)$ on $[\tilde{\eta},\tilde{T}+\tilde{\eta}] \cap [\tilde{\eta},T(\varphi))$.

This procedure can be continued to extend $\tilde{u}$ on $[\tilde{\eta},T(\varphi))$; this is possible since for every interval of existence $[\tilde{\eta},\tilde{T}+\tilde{\eta}]$ we get by (III.3.31) that

$$\tilde{u} = u \in C^{o}([\tilde{\eta},\tilde{T}+\tilde{\eta}] \cap [\tilde{\eta},T(\varphi)), H_p(\Omega))$$

which gives the a-priori estimate for $\|\tilde{u}(t)\|_{H_p(\Omega)}$, $\tilde{\eta} \le t < T(\varphi)$, being necessary to continue $\tilde{u}$ on $[\tilde{\eta},T(\varphi))$ according to theorem III.3.4. □

Finally we want consider an application of theorem II.3.10 to the equations of Navier-Stokes. It deals with the question to find local solutions of the Navier-Stokes equations being as smooth as possible at the initial time $t=0$ without any compatibility condition of an order higher than φ has to assume boundary values 0. Our result is as follows:

<u>Theorem III.3.7:</u> <u>Let</u> $p > \max(1,\frac{n}{3})$. <u>Let</u> $\varphi \in D(A^{1+\frac{1-\varepsilon}{2p}})$ <u>for any</u> $\varepsilon \in (0,1)$. <u>Let</u>

$$f \in C^{o}([0,T],(L^{p}(\Omega))^{n}) \text{ for all } T>0,$$

$$\int_0^t e^{-(t-s)A_p} P_p f(s)\, ds \in D(A^{1+\frac{1-\varepsilon}{2p}}),\ 0 \le t,$$

$$A_p^{1+\frac{1-\varepsilon}{2p}} \int_0^{\cdot} e^{-(\cdot-s)A_p} P_p f(s)\, ds \in C^{o}([0,T],H_p(\Omega)) \text{ for all } T>0.$$

<u>Then there is a</u> $T(\varphi)$, $0 < T(\varphi) \le +\infty$, <u>with the following property</u>: <u>There is one and only one</u> u <u>with</u>

$$u \in C^1([0,T(\varphi)),H_p(\Omega)),$$

$$u(t) \in D(A^{1+\frac{1-\varepsilon}{2p}}),$$

$$A_p^{1+\frac{1-\varepsilon}{2p}} u(.) \in C^0([0,T(\varphi)),H_p(\Omega)),$$

$$u' + A_p u + M(u) = Pf,$$
$$u(0) = \varphi.$$

_If_ $T(\varphi) < +\infty$ _then_

$$\lim_{t \uparrow T(\varphi)} \|A_p^{1-2\varepsilon'} u(t)\| = +\infty$$

for some $\varepsilon' \in (0,\frac{1}{2})$. $T(\varphi)$ _is the quantity from theorem_ III.3.1.

Proof: For the proof we omit the index p. Also the norm of $H_p(\Omega)$ is denoted by $\|.\|$. We want to determine κ,ρ_1,ρ_2 in order to apply theorem II.3.10.

We claim that $u\cdot\nabla u \in A(D(B^{\frac{3}{2}}),(\overset{\circ}{H}{}^{1,p}(\Omega))^n)$. Deriving formally we get

$$\frac{\partial}{\partial x_i}(u\cdot\nabla u) = \frac{\partial u}{\partial x_i}\cdot\nabla u + u\cdot\nabla\frac{\partial u}{\partial x_i}.$$

Estimating the term on the right hand side, we have by Sobolev and the remark after theorem III.2.5

$$\|\frac{\partial u}{\partial x_i}\cdot\nabla u\|_{(L^p(\Omega))^n} \le c\|u\|_{(H^{3-\gamma,p}(\Omega))^n}\|u\|_{(H^{3-\gamma,p}(\Omega))^n},$$

$$\le c\|B^{\frac{3}{2}}u\|^2,$$

$$\|u\cdot\nabla\frac{\partial u}{\partial x_i}\|_{(L^p(\Omega))^n} \le \|u\|_{(C^0(\overline{\Omega}))^n}\|u\|_{(H^{2,p}(\Omega))^n} \le c\|B^{\frac{3}{2}}u\|^2,$$

where the proof is done in a way similar to the proof of theorem III.3.1. Now a corresponding Lipschitz estimate can be derived in the same way. Thus $u\cdot\nabla u \in (\overset{\circ}{H}{}^{1,p}(\Omega))^n$ and the derivatives can be gained by formal differentiation. Then it can be shown as in the proof of theorem III.3.3 that $u\cdot\nabla u \in D(A(B^{\frac{3}{2}}),(\overset{\circ}{H}{}^{1,p}(\Omega))^n)$. An inspection of the Hölder exponents in the proof of theorem III.3.1 even gives with theorem III.2.6 that

$$\text{(III.3.32)}\quad u\cdot\nabla u \in A(D(B^{1-2\varepsilon'}),L^p(\Omega))$$

for some $\varepsilon' \in (0,\frac{1}{2})$. Considering the remark after theorem III.2.5 on the domains of B^ρ we get

$$\text{(III.3.33)}\quad u\cdot\nabla u \in A(D(B^{\frac{3}{2}}),D(B^{\frac{1}{2}})).$$

Thus by theorem II.2.1

$$u\cdot\nabla u \in A(D(A^{\gamma+\eta}),D(A^{\delta-\eta}))$$

for any $\eta > 0$ with

$$\gamma = \gamma(\sigma) = \sigma(1-2\varepsilon') + (1-\sigma)\frac{3}{2},$$

$$\delta = \delta(\sigma) = (1-\sigma)\frac{1}{2},$$

$$\gamma(\sigma)+\eta \leq \frac{3}{2}.$$

Let

$$\delta(\sigma)-\eta = \frac{1}{2}(1-\sigma)-\eta = \frac{1-\varepsilon}{2p}, \text{ thus}$$

$$1-\sigma = \frac{1-\varepsilon}{2p}+2\eta, \quad \sigma = 1 - \frac{1-\varepsilon}{2p} - 2 .$$

Then

$$\gamma(\sigma) = 1 + \frac{1-\varepsilon}{2p} + \frac{\varepsilon'}{p} + \eta + 2\eta\varepsilon' - \varepsilon',$$

$$\gamma(\sigma)+\eta = 1 + \frac{1-\varepsilon}{2p} + \frac{\varepsilon'}{p} + 2\eta(1+\varepsilon') - \varepsilon'.$$

If η is chosen sufficiently small then

$$\gamma(\sigma)+\eta < 1+\frac{1-\varepsilon}{2p}.$$

Let $\kappa = 1+\frac{1-\varepsilon}{2p}+\frac{\varepsilon'}{p}+2\eta(1+\varepsilon')-\varepsilon'$, $\kappa-(1-\rho_2) = \frac{1-\varepsilon}{2p}$, thus $\kappa+\rho_2 = 1+\frac{1-\varepsilon}{2p}$, $\kappa-(1-\rho_1) = 1-\varepsilon'$, $\frac{1-\varepsilon}{2p}+\frac{\varepsilon'}{p}+2\eta(1+\varepsilon')-\varepsilon'+\rho_1 = 1-\varepsilon'$, $\rho_1 = 1-(\frac{1-\varepsilon}{2p}+\frac{\varepsilon'}{p}+2\eta(1+\varepsilon'))$. Consequently

$$P(u\cdot\nabla u) \in A(D(A^{\kappa-(1-\rho_1)}),H_p(\Omega)),$$

$$u\cdot\nabla u \in A(D(A^{\kappa}),D(B^{\kappa-(1-\rho_2)})).$$

Employing Giga's result for the fractional powers of A we get $D(A^{\kappa-(1-\rho_2)}) = (\overset{o}{H}{}^{\frac{1-\varepsilon}{p},p}(\Omega))^n \cap H_p(\Omega) = (H^{\frac{1-\varepsilon}{p},p}(\Omega))^n \cap H_p(\Omega)$, by theorem III.2.6 we have

$$PD(B^{\kappa-(1-\rho_2)}) = P(\overset{o}{H}{}^{\frac{1-\varepsilon}{p},p})^n,$$

$$\subset \overline{H_p(\Omega) \cap \{v|v \in (C_o^\infty(\Omega))^n\}}^{\|\cdot\|_{(H^{\frac{1-\varepsilon}{p},p}(\Omega))^n}},$$

$$= H_p(\Omega) \cap D(B^{\kappa-(1-\rho_2)}) = D(A^{\kappa-(1-\rho_2)}).$$

Theorem II.3.10 completes the proof. □

§ 4. Global Existence for Small Data. Extension of the Previous Results to Arbitrary Dimensions

First we deal with the question if the local solutions constructed in theorems III.3.1,4,5,6 exist for all times provided the initial values φ are small in appropriate norms and provided f is small for large times in a sense to be made precise later on. Our results are collected in the following theorem:

Theorem III.4.1: 1. Let $p > \frac{n}{3}, 1$, $p \neq \frac{3}{2}$. Let f be as in theorem III.3.1 and, let $Pf \in L^{\infty}((0,\infty), H_p(\Omega))$, $(Pf)' \in L^{\infty}((0,\infty), H_p(\Omega))$, where by $(Pf)'$ we mean the derivative of Pf in the sense of lemma II.1.1. Let $\varphi \in D(A_p)$, let $A_p\varphi$ and f be small enough, i.e.:

$$\|P_p f(t)\|_{H_p(\Omega)} + \|P_p f'(t)\|_{H_p(\Omega)} + \|A_p \varphi\|_{H_p(\Omega)} \leq \varepsilon_1, 0 < t,$$

where ε_1 is a certain positive number depending on A_p. Then $T(\varphi) = +\infty$ in theorem III.3.1.

2. Let $p > n$. Let f be as in theorem III.3.4 and, additionally, let $Pf \in L^{\infty}((0,\infty), H_p(\Omega))$. Let $\varphi \in H_p(\Omega)$, let $\|P_p f(t)\|_{H_p(\Omega)} + \|\varphi\|_{H_p(\Omega)}$ be small enough, i.e.

$$\|P_p f(t)\| + \|\varphi\|_{H_p(\Omega)} \leq \varepsilon_2, \ 0 < t$$

where ε_2 is a certain positive number depending on A_p. Then $T(\varphi) = +\infty$ in theorem III.3.4.

3. Let $p=2$. Let f be as in theorem III.3.5 and, additionally, let $P_2 f \in L^\infty((0,\infty),H_2(\Omega))$. Let $\varphi \in D(A_2^{1/4})$, let $\|P_2 f(t)\|_{H_2(\Omega)} + \|A_2^{1/4}\varphi\|_{H_2(\Omega)}$ be small enough in the sense above. Then $T(\varphi)=+\infty$ in theorem III.3.5.

4. Let $p=n$. Let f be as in theorem III.3.6 and, additionally, let $f \in L^\infty((0,\infty),H_n(\Omega))$. Let $\varphi \in H_n(\Omega)$, let $\|P_n f(t)\|_{H_n(\Omega)} + \|\varphi\|_{H_n(\Omega)}$ be small enough in the sense above. Then $T(\varphi)=+\infty$ in theorem III.3.6.

Proof: The key for our considerations is the estimate

$$\|e^{-tA_p}\|_{H_p(\Omega)} \le ce^{-\delta t}, \quad p>1,\ p \neq \tfrac{3}{2},$$

which was proved in III.1 (see (III.1.15)). As for the first case we remark that $(\|\cdot\| = \|\cdot\|_{H_p(\Omega)})$

$$\|M(u(t))\| \le c\|Au(t)\| \, \|A^{1-\rho}u(t)\|,$$
$$\|M'(u(t))\| \le c\|Au(t)\| \, \|A^{1-\rho}u'(t)\|,$$

as it follows from lemma II.1.1; here u is the solution constructed in theorem III.3.1. From (II.1.4) and (II.1.5) we get the estimates

$$\|Au(t)\| \le ce^{-\delta t}(\|A\varphi\| + \|M(\varphi)\| + \|f(0)\|) + c\|Au(t)\| \, \|A^{1-\rho}u(t)\| + \\ + \|f(t)\| + \int_0^t ce^{-\delta(t-s)}(\|Au(s)\| \, \|A^{1-\rho}u'(s)\| + \\ \|f'(s)\|)\, ds,$$

$$\|A^{1-\rho}u'(t)\| \le \frac{ce^{-\delta t}}{t^{1-\rho}}(\|A\varphi\| + \|M(\varphi)\| + \|f(0)\|) + \\ + \int_0^t \frac{ce^{-\delta(t-s)}}{(t-s)^{1-\rho}}(\|Au(s)\| \, \|A^{1-\rho}u'(s)\| + \|f'(s)\|)\, ds.$$

Let us consider the system of integral equations

$$v_1(t) = ce^{-\delta t}(\|A\varphi\|+\|M(\varphi)\| +\|f(0)\|) + \|f(t)\| + $$
$$+ cv_1^2(t) + \int_0^t ce^{-\delta(t-s)}(v_1(s)v_2(s) +\|f'(s)\|)\, ds,$$

$$v_2(t) = \frac{ce^{-\delta t}}{t^{1-\rho}}(\|A\varphi\|+\|M(\varphi)\| +\|f(0)\|) + $$
$$+ \int_0^t \frac{ce^{-\delta(t-s)}}{(t-s)^{1-\rho}}(v_1(s)v_2(s) +\|f'(s)\|)\, ds.$$

We set $\varepsilon_o = c_1(\|A\varphi\|+\|M(\varphi)\| +\|f(0)\|) + c_2 \sup\limits_{0\le t<+\infty} (\|f(t)\| +\|f'(t)\|)$. Then it is easily seen that the system above has a unique solution within the class of functions $v_1 \in C^o([0,\infty))$, $0 \le v_1(t) \le 2\varepsilon_o$, and $v_2 \in C^o((0,\infty))$, $0 \le v_2(t) \le \frac{2\varepsilon_o}{t^{1-\rho}} + 2\varepsilon_o$, if c_1, c_2 are properly adjusted and ε_o is sufficiently small. Since

$$\|Au(t)\| \le v_1(t),$$
$$\|A^{1-\rho}u'(t)\| \le v_2(t),$$

the quantity $\|Au(t)\|$ is estimated a-priori. Thus, the first part of our theorem is proved.

As for the second part we have to consider the integral inequalities $(\|\cdot\|=\|\cdot\|_{H_p(\Omega)})$

$$\|A_p^{1-\rho}u(t)\| \le \frac{ce^{-\delta t}}{t^{1-\rho}}\|\varphi\| + \int_0^t \frac{ce^{-\delta(t-s)}}{(t-s)^{1-\rho}}(\|A_p^{1-\rho}u(s)\|\cdot\|u(s)\| + \|P_pf(s)\|)\, ds,$$

$$\|u(t)\| \le ce^{-\delta t}\|\varphi\| + \int_0^t ce^{-\delta(t-s)}(\|A_p^{1-\rho}u(s)\| \ \|u(s)\| +\|P_pf(s)\|)\, ds.$$

Now we again consider a system of integral equations

$$v_1(t) = \frac{ce^{-\delta t}}{t^{1-\rho}}\|\varphi\| + \int_0^t \frac{ce^{-\delta(t-s)}}{(t-s)^{1-\rho}}(v_1(s)v_2(s) + \|P_p f(s)\|)\,ds,$$

$$v_2(t) = ce^{-\delta t}\|\varphi\| + \int_0^t ce^{-\delta(t-s)}(v_1(s)v_2(s) + \|P_p f(s)\|)\,ds.$$

In order to prove the second part of our theorem this system is treated in the same way as the system before.

As for the third case observe that we have the estimate $(\|\cdot\| = \|\cdot\|_{H_2(\Omega)})$

$$\|A_2^{\frac{3}{4}}u(t)\| \le \frac{ce^{-\delta t}}{t^{1/2}}\|A_2^{\frac{1}{4}}\varphi\| + \int_0^t \frac{ce^{-\delta(t-s)}}{(t-s)^{3/4}}(\|A_2^{\frac{1}{4}}u(s)\|^{\frac{1}{2}} \cdot \|A_2^{\frac{3}{4}}u(s)\|^{\frac{3}{2}} + \|P_2 f(s)\|)\,ds,$$

$$\|A_2^{\frac{1}{4}}u(t)\| \le ce^{-\delta t}\|A_2^{\frac{1}{4}}\varphi\| + \int_0^t \frac{ce^{-\delta(t-s)}}{(t-s)^{1/4}}(\|A^{\frac{1}{4}}u(s)\|^{\frac{1}{2}} \cdot \|A_2^{\frac{3}{4}}u(s)\|^{\frac{3}{2}} + \|P_2 f(s)\|)\,ds,$$

and we see that the further procedure is very much like that one of the foregoing cases. It gives an a-priori estimate for $\|A_2^{\frac{3}{4}}u(t)\|$ which is sufficient for $T(\varphi) = +\infty$ (see theorem II.3.6).

As for the fourth case $(\|\cdot\| = \|\cdot\|_{H_n(\Omega)})$ we have the estimates

$$\|A_n^{\frac{1}{4}}u(t)\| \le ce^{-\delta t}\|\varphi\| + \int_0^t \frac{ce^{-\delta(t-s)}}{(t-s)^{3/4}}(\|A_n^{\frac{1}{4}}u(s)\|^2 + \|A_n^{-\frac{1}{2}}P_n f(s)\|)\,ds.$$

As before this yields an a-priori estimate for $\|A_n^{\frac{1}{4}}u(t)\|$. This means that we have estimated a-priori $\|u(t)\|_{H_{n+\delta}(\Omega)}$ for some

$\delta > 0$. Identifying for $t > 0$ the solution in theorem III.3.6 with that in theorem III.3.4 (as it was done in fact in the proof of theorem III.3.6) and applying the part of theorem III.3.4 which concerns the case $T(\varphi) < +\infty$ we see that $T(\varphi) = +\infty$. Thus our theorem is proved. □

Now we want to give some attention to the question which of our results on the local solvability of the Navier-Stokes equations remain true for $n \geq 4$. This comes out to the question if the abstract setting of II.3 is still applicable. Since much work has already been done in this direction it turns out that most of our results remain true wordly for $n \geq 4$. Observe that we always have used the symbol n for the dimension, although we have restricted ourselves to $n = 2,3$. This allows us to see better where a generalization is possible.

We start with the one exceptional case, namely theorem III.3.5. This theorem is strictly confined to $n = 3$ as it is clear from the proof where we have made strong use of the assumption $n = 3$.

All other cases can be generalized to arbitrary $n \geq 4$. The reasons for this are as follows: First of all Solonnikov's potential theoretical estimates can be carried over to arbitrary dimensions. This was done in [W4],[W2] for $p \geq 2$ and $\varphi \in H^{2,p}(\Omega) \cap H_p(\Omega)$, but it also works for $p > 1$, $\varphi \in H^{2,p}(\Omega) \cap H_p(\Omega)$. If the initial values are chosen from $H^{2,p}(\Omega) \cap H_p(\Omega)$ the exception: $p \neq \frac{3}{2}$ is not needed in Solonnikov's estimates as was already pointed out after theorem III.1.3. This choice is also sufficient for the proof of the resolvent estimates for $p > 1$ (see [W2] for $p \geq 2$); however, in order to deal with all $p > 1$, $n \geq 4$ some modifications in the proof are necessary, caused e.g. by the fact that we have used the estimate $\|u\|_{C^0(\overline{\Omega})} \leq c\|A_2 u\|$, $u \in D(A_2)$, which is valid only in three dimensions. These modifications can be carried through on the basis of the linear theory for (III.1.8) since (III.1.8) is elliptic as was mentioned before.

On the other hand Giga [Gi 2] has proved the resolvent estimates in question for any space dimension. Let us remark again in this connection that all results on the projections P_p that have been mentioned in the auxiliary propositions are valid for arbitrary $n \geq 2$.

As a consequence of the foregoing considerations the results of III.2 can be carried over to $n \geq 4$ without further change. This in particular means that theorems III.3.1, III.3.2, III.3.3, proposition III.3.1, theorems III.3.4, III.3.6, III.3.7, and III.4.1, 1.,2.,3. can be carried over to $n \geq 4$ immediately with the exception of one minor point: For the proof of

(III.4.1) $\mathrm{Re}(P_q(u(t)\cdot\nabla v),v) = 0$

we have used the chain rule for $\nabla(v_\lambda \overline{v}_\lambda)$ in that way that we have worked in $H^{2,q}(\Omega)$, $q \geq 2 > \frac{n}{q}$ for $n = 3,2$; the reader can easily show by himself that (III.4.1) remains true also in the case $n \geq 4$ if $u \in C^0([0,T],H_{n+\delta}(\Omega))$ for some $\delta > 0$ and $v \in H^{2,q}(\Omega) \cap \overset{\circ}{H}{}^{1,q}(\Omega) \cap H_q(\Omega)$, $q \geq 2$.

§ 5. Comments to Chapter III

Our comments may be rather brief at this point since most has been included in § 4, when the case of higher dimensions was considered.

<u>To § 1:</u> The method of introducing in (III.1.2) a coefficient independent from T was given in [W5]. $\|u\|_{(L^p(\Omega))^n}$, $\|\pi\|_{L^p(\Omega)}$ on the right hand side of (III.1.9) may probably be removed by the use of the theory of elliptic systems in [ADN II] if one knows

that u is unique and π is unique up to a constant (In [Tem,p. 33] this is stated, but without proof). Here we can rely on [L].

To § 2: The material here has appeared in [W 2] (For any number of space dimensions). There is an estimate analogous to (III.1.2) in the $C^{\alpha}(\overline{\Omega})$-spaces; this estimate is of the type being known for linear parabolic equations (see [LUS]). As a consequence one can derive an estimate for the resolvent of the Stokes operator A in $C^{\alpha}(\overline{\Omega})$, namely

$$\text{(III.5.1)} \quad \|(\lambda+A)^{-1}u\|_{C^{\alpha}(\overline{\Omega})} \leq \frac{c}{|\lambda|^{1-\alpha/2}},$$

in a way analogous to III.1. Therefore A generates a semigroup e^{-tA} in the divergence free part of $C^{\alpha}(\overline{\Omega})$ which is not strongly continuous for $t \to 0$. The corresponding result for elliptic operators is well known (cf. [W6]). Bemelmans [Be] has given a different proof for the resolvent estimate just mentioned; this proof even works in unbounded domains.

To § 3: As Bemelmans [Be] has proved one can build up a local existence theory for the Navier-Stokes equations for bounded and unbounded domains Ω making use of (III.5.1). For unbounded domains this in particular yields solutions which behave for $|x| \to \infty$ in a way being different from $L^p(\Omega)$-solutions.

The abstract setting for dealing with bad initial values was given already in [W7] and, within a more general framework, in [W2] (including $H_n(\Omega)$ initial values; [W2] also contains a proof of the local strong existence for the Cauchy problem with $H_n(\mathbb{R}^n)$ initial values). Proofs being somewhat different for the local strong existence with $H_n(\Omega)$ initial values were given by Weissler [We] and Giga and Miyakawa [GiM]. A completely different proof of this theorem can be found in [SoW 1] where Solonnikov's estimates are used as a basic tool; this last device for proving local existence for the Navier-Stokes equations goes back to Sohr.

Theorem III.3.7 was proved first by Rautmann [R] in the case $p=2$, $n=3$. It is clear from the Navier-Stokes equations that a compatibility condition higher than $\varphi=0$ on $\partial\Omega$ may involve a boundary condition on the pressure π being difficult to check. Rautmann [R] has given a more detailed description of the situation occurring here. He also points out that for an efficient numerical treatment it would be desirable to work with solutions being as smooth as possible at least at $t=0$.

IV. Global Existence and Global Regularity for the Navier-Stokes Equations

It has been known since 1951 that the Navier-Stokes equations

$$\frac{\partial u}{\partial t} - \nu\Delta u + u\cdot\nabla u + \nabla\pi = f$$
$$\nabla\cdot u = 0$$

over $(0,T)\times\Omega$ with boundary conditions $u(0) = \varphi$, $u|\partial\Omega = 0$ can be solved weakly for all times. This result is due to E. Hopf [Ho]. It is the aim of this chapter to study the connections between both types of solutions, the weak ones and the strong ones. Up to now it is neither known if the weak solutions are unique nor if they are regular. Unless restrictions are made explicitly the space dimension n can be chosen also arbitrary in this chapter, if one uses the n-dimensional analogues of the linear theory (cf. for this point III.4).

§ 1. Weak solutions

First we want to give the notion of a weak solution.

Definition IV.1.1: Let $\varphi \in H_2(\Omega)$, $f \in L^2((0,T),(H^{-1,2}(\Omega))^n)$. An element

$$u \in L^\infty((0,T),(L^2(\Omega))^n) \cap L^2((0,T),(\overset{\circ}{H}{}^{1,2}(\Omega))^n)$$

which fulfills

$$\nabla\cdot u(t) = 0$$

for almost all $t \in (0,T)$ and which is weakly continuous in $(L^2(\Omega))^n$ from $[0,T]$ to $(L^2(\Omega))^n$ is called a weak solution of the Navier-

Stokes equations

$$\text{(IV.1.1)}\quad \frac{\partial u}{\partial t} - \nu\Delta u + u\cdot\nabla u + \nabla\pi = f$$
$$\nabla\cdot u = 0$$

over $(0,T)\times\Omega$ with boundary values $u(0)=\varphi$, $u|\partial\Omega = 0$ and inhomogeneous term f [1] if

$$\text{(IV.1.2)}\quad -\int_0^T (u,\psi')\,dt + \nu\int_0^T (\nabla u,\nabla\psi)\,dt - \int_0^T (u\cdot u,\nabla\psi)\,dt =$$
$$= (\varphi,\psi(0)) + \int_0^T (f,\psi)\,dt$$

for all testing functions $\psi \in C^{0,1}([0,T],(L^2(\Omega))^n)\cap C^0([0,T],(\overset{o}{H}{}^{1,2}(\Omega))^n)\cap L^\infty((0,T),(\overset{o}{H}{}^{1,\frac{n}{2}}(\Omega))^n)$ with $\nabla\cdot\psi(t)=0$ on $[0,T]$, $\psi(T)=0$

n is now an arbitrary number ≥ 2. Ω is a bounded domain of $\mathbb{R}^n$ of class C^∞. First we want to discuss the new notion. The statement $u(0)=\varphi$ makes sense since u is supposed to be weakly continuous in $(L^2(\Omega))^n$. Since

$$\nabla\cdot u(t) = 0 \text{ a.e. in } (0,T),$$

$$u(t)\in(\overset{o}{H}{}^{1,2}(\Omega))^n \text{ a.e. in } (0,T)$$

and since $H_*^{1,2}(\Omega) = \{v\,|\,v\in(\overset{o}{H}{}^{1,2}(\Omega))^n,\ \nabla\cdot v = 0\}$ is the closure of the vector fields in $(C_0^\infty(\Omega))^n$ with vanishing divergence with respect to the $(H^{1,2}(\Omega))^n$-norm, we see that

$$u(t)\in H_2(\Omega) \text{ a.e. in } (0,T).$$

$H_2(\Omega)$ being closed against weak convergence in $(L^2(\Omega))^n$ it therefore follows that

---

[1] In refering to this definition later on there is frequently none or only partial mentioning of the data φ,f if no confusion can arise.

$u(t) \in H_2(\Omega)$ on $[0,T]$.

Thus the assumption $u(0) \in H_2(\Omega)$ is no additional requirement. We turn to the question if (IV.1.2) does make any sense. This is clear as it concerns the first two members of (IV.1.2). $u \cdot u$ is the vector with components $u_i \cdot u_l$, $1 \le i,l \le n$, where u_i are the components of u; if u, ψ have some additional degree of regularity then we already know that

$$(IV.1.3) \quad (u \cdot \nabla u, \psi) = -(u \cdot u, \nabla \psi)$$

(see the proof of theorem III.3.6). By Hölder's inequality it follows that

$$(IV.1.4) \quad |(u \cdot u, \nabla\psi)| \le c\|u \cdot u\|_{(L^2(\Omega))^{n^2}} \|\nabla\psi\|_{(L^2(\Omega))^{n^2}}$$

$$\le c\|u\|^2_{(L^4(\Omega))^n} \|\nabla\psi\|_{(L^2(\Omega))^{n^2}},$$

or

$$(IV.1.5) \quad |(u \cdot u, \nabla\psi)| \le c\|u\|^2_{(L^{2n/(n-2)}(\Omega))^n} \|\nabla\psi\|_{(L^{n/2}(\Omega))^{n^2}} \text{ for } n \ge 3,$$

$$\le c\|u\|^2_{(L^4(\Omega))^n} \|\nabla\psi\|_{(L^2(\Omega))^{n^2}} \text{ for } n = 2.$$

Thus we see that all terms occurring in the definition of a weak solution are well defined, and moreover that it may be not allowed to insert u as a testing function (with the exception of the case $n = 2$ on which will come back later on).

Because of $(\nabla\pi, \psi) = 0$ for a classical solution it's easily seen that (IV.1.2) is fulfilled for a classical solution of the Navier-Stokes equations (we will not make this notion precise here since we do not need it). This raises the question where the pressure π has disappeared in our notion of a weak solution. In answering this question we need some preparations. From the definition of u it follows that

(IV.1.6) $-\nu\Delta u \in L^2((0,T),H^{-1,2}(\Omega))$,

from (IV.1.5) we get

(IV.1.7) $u\cdot\nabla u \in L^1((0,T),(H^{-1,\frac{n}{n-2}}(\Omega))^n)$ for $n=3$,

$u\cdot\nabla u \in L^1((0,T),H^{-1,2}(\Omega))$ for $n=2$.

By $H_*^{-1,p}(\Omega)$ we denote the dual of $(\overset{o}{H}{}_*^{1,q}(\Omega))^n = \{u|u \in (\overset{o}{H}{}^{1,q}(\Omega))^n, \nabla\cdot u=0\}$ endowed with the norm of $\overset{o}{H}{}^{1,q}(\Omega)$, $\frac{1}{p}+\frac{1}{q}=1$, $+\infty > p,q > 1$. Observe that

$$H_*^{-1,p}(\Omega) \supset (H^{-1,p}(\Omega))^n$$

with a continuous imbedding. Thus (IV.1.2) furnishes

(IV.1.8) $u' - \nu\Delta u + u\cdot\nabla u = f$

as an equation in $L^1((0,T),H_*^{-1,\frac{n}{n-2}}(\Omega)) + L^2((0,T),H_*^{-1,2}(\Omega))$ or in $L^1((0,T),H_*^{-1,2}(\Omega))$ ($n=3$, 2 resp.); $-\nu\Delta u$, f are simply defined by restriction. Thus

(IV.1.9) $u \in C^0([0,T],H_*^{-1,\frac{n}{n-2}}(\Omega)) + C^{\frac{1}{2}}([0,T],H_*^{-1,2}(\Omega))$

Since with a continuous imbedding

$$H_*^{-1,\frac{n}{n-2}}(\Omega) \subset H_*^{-1,2}(\Omega)$$

we get in particular for $n=3,2$ that

(IV.1.9) $u \in C^0([0,T],H_*^{-1,2}(\Omega))$.

Together with $u \in L^\infty((0,T),(L^2(\Omega))^n)$ it follows that u is weakly continuous from [0,T] into $(L^2(\Omega))^n$; thus our assumption on the weak continuity of u is redundant, but this needed a proof. Since we need some continuity property of u to explain in which sense u assumes its initial value we had nevertheless included the weak continuity of u in the definition IV.1.1.

We want to make a remark on dimensions n higher than 3 at this stage of our discussion: The definition IV.1.1 has already been given for $n \geq 2$ (compare [Li,1.6]; it's the same with (IV.1.5)). Since for $n \geq 4$ we have $H_*^{-1,\frac{n-2}{n}}(\Omega) \supset H_*^{-1,2}(\Omega)$ with a continuous imbedding we get

$$\text{(IV.1.10)} \quad u \in C^0([0,T], H_*^{-1,\frac{n-2}{n}}(\Omega))$$

and together with $u \in L^\infty((0,T),(L^2(\Omega))^n)$ the weak continuity of u.

We want now to turn to the question after the pressure π. We follow the presentation of [Li, 1.6, p. 69]. Now considering u', $-\nu\Delta u$, $u\cdot\nabla u$, f as time dependent distributions, where $u\cdot\nabla u$ is defined by (IV.1.3) and where $\mathcal{D}'(.)$ denotes the space of distributions over an open set of $\mathbb{R}^n$ or $\mathbb{R}^{n+1}$), we get

$$(\mathcal{D}'(\Omega))^n \ni S(t) = u'(t) - \nu\Delta u(t) + u\cdot\nabla u(t) - f(t).$$

Then $S \in (\mathcal{D}'((0,T)\times\Omega))^n$. We want to characterise S. Let $\psi \in (C_0^\infty(\Omega))^n$, $\nabla\cdot\psi = 0$. Let $0 < \tilde{T} < T$, $0 < 2\varepsilon < \tilde{T}$,

$$\zeta_\varepsilon^{\tilde{T}}(t) = \begin{cases} \frac{1}{\varepsilon}t, & 0 \leq t \leq \varepsilon, \\ 1, & \varepsilon \leq t \leq \tilde{T}-\varepsilon, \\ -\frac{1}{\varepsilon}t + \frac{1}{\varepsilon}\tilde{T}, & T-\varepsilon \leq t \leq \tilde{T}, \\ 0 & \text{otherwise.} \end{cases}$$

Inserting $\zeta_\varepsilon^{\tilde{T}}(t)\psi$ as a testing function in (IV.1.2) we get

$$-\frac{1}{\varepsilon}\int_0^\varepsilon (u,\psi)\,dt + \frac{1}{\varepsilon}\int_{\tilde{T}-\varepsilon}^{\tilde{T}} (u,\psi)\,dt + \nu\int_0^{\tilde{T}} (\nabla u, \zeta_\varepsilon^{\tilde{T}}\nabla\psi)\,dt -$$

$$-\int_0^{\tilde{T}} (u\cdot u, \zeta_\varepsilon^{\tilde{T}}\nabla\psi)\,dt = \int_0^{\tilde{T}} (f, \zeta_\varepsilon^{\tilde{T}}\psi)\,dt.$$

Letting ε tend to 0 the weak continuity of u shows that

$$- (\varphi,\psi) + (u(\tilde{T}),\psi) + \nu\int_0^{\tilde{T}} (\nabla u,\nabla\psi)\,dt -$$

$$- \int_0^{\tilde{T}} (u\cdot u,\nabla\psi)\,dt = \int_0^{\tilde{T}} (f,\psi)\,dt.$$

Now $\psi \in L^\infty((0,T),(\overset{\circ}{H}{}^{1,n/2}(\Omega))^n) \cap L^2((0,T),(\overset{\circ}{H}{}^{1,2}(\Omega))^n)$. Since $u' \in$ $\in L^1((0,T),H_*^{-1,\frac{n-2}{n}}(\Omega)) + L^2((0,T),H_*^{-1,2}(\Omega))$, i.e.

$$u' \in L^1((0,T),H_*^{-1,2}(\Omega)),\ n = 2,3,$$

$$u' \in L^1((0,T),H_*^{-1,\frac{n-2}{n}}(\Omega)),$$

we arrive at

$$\int_0^{\tilde{T}} (u'(t),\psi)\,dt = (u(\tilde{T}),\psi) - (\varphi,\psi)$$

and consequently

$$\int_0^{\tilde{T}} (u'(t),\psi)\,dt + \nu\int_0^{\tilde{T}} (\nabla u(t),\nabla\psi)\,dt - \int_0^{\tilde{T}} (u\cdot u(t),\nabla\psi)\,dt$$

$$= \int_0^{\tilde{T}} (f(t),\psi)\,dt.$$

Now we can differentiate this equation with respect to $\tilde{T}$. This gives

$$(S(t),\psi) = 0 \text{ a.e. on } (0,T)$$

since $\tilde{T}$ was arbitrary. According to [Li,1.6, pp. 67-69] there is a $\pi \in \mathcal{D}'((0,T)\times\Omega)$ with $\pi(t) \in \mathcal{D}'(\Omega)$ a.e. on $(0,T)$ and

$$S(t) = -\nabla\pi(t).$$

π is called the pressure, better "a pressure", since π is not unique; we do not want to go into details here.

If $0 < t_o < \tilde{T} < T$ then one can also define the notion of a weak solution of the Navier-Stokes equations (IV.1.1) over $(t_o,\tilde{T})\times\Omega$ with boundary values $u(t_o) = \varphi$, $u|\partial\Omega = 0$. Since it is clear how we shall do that, we want to omit the definition and use this notion freely in the sequel.

As one expects the following theorem holds:

Theorem IV.1.1: Let $0 \le t_o < \tilde{T} \le T$. Let u be a weak solution of (IV.1.1) over $(0,T)\times\Omega$ with boundary values $u(0) = \varphi$, $u|\partial\Omega = 0$. Then the restriction of u to $(t_o,\tilde{T})$ is also a weak solution of (IV.1.1) over $(t_o,\tilde{T})\times\Omega$ with boundary values $u(t_o)$, $u|\partial\Omega = 0$.

Proof: Let $0 < t_o < \tilde{T} < T$. Let ψ be a testing function with

$$\psi \in C^{0,1}([t_o,\tilde{T}],(L^2(\Omega))^n),$$

$$\psi \in C^0([t_o,\tilde{T}],(\overset{o}{H}{}^{1,2}(\Omega))^n) \cap L^\infty((t_o,\tilde{T}),(\overset{o}{H}{}^{1,\frac{n}{2}}(\Omega))^n),$$

$$\psi(\tilde{T}) = 0.$$

Let $0 < \varepsilon < t_o$, $\tilde{T}-t_o$. We first continue ψ to $[\tilde{T},T]$ by setting

$$\tilde{\psi}_\varepsilon(t) = 0, \quad \tilde{T} \le t \le T$$

and then to $[0,t_o]$ by setting

$$\tilde{\psi}_\varepsilon(t) = 0, \quad 0 \le t \le t_o,$$

$$\tilde{\psi}_\varepsilon(t) = \frac{1}{\varepsilon}(t-t_o)\psi(t), \quad t_o \le t \le t_o+\varepsilon,$$

$$\tilde{\psi}_\varepsilon(t) = 1, \quad t_o+\varepsilon \le t \le \tilde{T}.$$

Thus we have constructed an admissible testing function for (IV.1.1) and by inserting ψ_ε into (IV.1.2)

$$\text{(IV.1.11)}\quad -\int_{t_o}^{\tilde T} (u,\tilde\psi_\varepsilon')\,dt + \int_{t_o}^{\tilde T} \nu(\nabla u,\nabla\tilde\psi_\varepsilon)\,dt - $$

$$-\int_{t_o}^{\tilde T} (u\cdot u,\nabla\tilde\psi_\varepsilon)\,dt = \int_0^T (f,\tilde\psi_\varepsilon)\,dt.$$

Now we have

$$\text{(IV.1.12)}\quad -\int_{t_o}^{\tilde T} (u,\tilde\psi_\varepsilon')\,dt = -\int_{t_o}^{t_o+\varepsilon} (u,\psi_\varepsilon')\,dt - \int_{t_o+\varepsilon}^{\tilde T} (u,\psi')\,dt$$

$$= -\int_{t_o}^{t_o+\varepsilon} (u(t),\tfrac{1}{\varepsilon}\psi(t))\,dt - \int_{t_o}^{t_o+\varepsilon} (u(t),\tfrac{1}{\varepsilon}(t-t_o)\psi'(t))\,dt$$

$$-\int_{t_o+\varepsilon}^{\tilde T} (u,\psi')\,dt.$$

Because of the weak continuity of u the first integral on the right of (IV.1.12) is converging to $-(u(t_o),\psi(t_o))$ for $\varepsilon\to 0$. The second integral is converging to 0 for $\varepsilon\to 0$, the third one to

$$-\int_{t_o}^{T} (u,\psi')\,dt$$

for $\varepsilon\to 0$. As the convergence properties for $\varepsilon\to 0$ of the remaining integrals in (IV.1.11) may be clear, we finally get

$$-\int_{t_o}^{\tilde T} (u,\psi')\,dt + \int_{t_o}^{\tilde T} \nu(\nabla u,\nabla\tilde\psi_\varepsilon)\,dt - \int_{t_o}^{\tilde T} (u\cdot u,\nabla\tilde\psi_\varepsilon)\,dt =$$

$$= (u(t_o),\psi(t_o)) + \int_{t_o}^{\tilde T} (f,\psi)\,dt.$$

The cases $0=t_o$ or $\tilde T=T$ can be treated analogously. Our theorem is proved. □

As was already mentioned, E. Hopf [Ho] has proved the existence of a weak solution for any $T > 0$ and in any dimension n. The proof of this important theorem is now part of the standard literature, cf. e.g. [Tem, III.3.1], [Li, 1.6]. Thus we do not feel the necessarity to give a proof here. Since the method of proof consists in the Galerkin approximation procedure, the weak solutions constructed in this way have an important additional property (in comparison with definition IV.1.1), namely they fulfill the energy inequality. It will be made clear in what follows what that means. We state the existence theorem for weak solutions as

Theorem IV.1.2: Let $T > 0$ be arbitrary. Let $\varphi \in H_2(\Omega)$, $f \in L^2((0,T), (H^{-1,2}(\Omega)^n)$ be real valued. Then there exists a real valued weak solution of the Navier-Stokes equations (IV.1.1) over $(0,T)\times\Omega$ with boundary values $u(0) = \varphi$, $u|\partial\Omega = 0$ and inhomogeneous term f which has the following additional properties:

(IV.1.13) u is continuous from the right in 0 in the $(L^2(\Omega))^n$-norm with respect to t,

the energy inequality holds, i.e.:

$$\text{(IV.1.14)} \quad \|u(t)\|^2_{(L^2(\Omega))^n} + 2\nu \int_s^t \|\nabla u(\sigma)\|^2_{(L^2(\Omega))^{n^2}} d\sigma$$

$$\leq \|u(s)\|^2_{(L^2(\Omega))^n} + 2 \int_s^t (f(\sigma), u(\sigma))\, d\sigma$$

for almost all $s \in (0,T)$ and $s = 0$, and all t, $s \leq t \leq T$.

If additionally

$$f \in \bigcap_{T>0} L^2((0,T), (H^{-1,2}(\Omega))^n)$$

then there exists a real weak solution u of the Navier-Stokes equations (IV.1.1) over $(0,+\infty)\times\Omega$ with boundary values $u(0)=\varphi$, $u|\partial\Omega=0$ by which we mean a weakly continuous mapping from $[0,+\infty)$ into $(L^2(\Omega))^n$ with:

$$u \in \bigcap_{T>0} L^\infty((0,T),(L^2(\Omega))^n) \cap \bigcap_{T>0} L^2((0,T),(\overset{\circ}{H}{}^{1,2}(\Omega))^n),$$

$\nabla\cdot u(t)=0$ for almost all $t\in(0,+\infty)$,

(IV.1.2) is fulfilled for every $T>0$,

(IV.1.13) holds,

(IV.1.14) holds for almost all $s\in(0,+\infty)$, $s=0$, and all t, $s\leq t<+\infty$.

(IV.1.14) for $s=0$ follows by (IV.1.13) and by (IV.1.14) for almost all $s\in(0,T)$ and all t, $s\leq t\leq T$. We want to make a few remarks concerning (IV.1.14). If u is a classical real solution then even (IV.1.14) holds with equality sign for all s,t. This is easily seen if one multiplies the Navier-Stokes equations scalarly by u since

$$(u\cdot\nabla u,u) = \sum_{l=1}^{n}\int_\Omega\sum_{i=1}^{n} u_i \frac{\partial u_l}{\partial x_i} u_l\, dx,$$

$$= \frac{1}{2}\sum_{l=1}^{n}\int_\Omega\sum_{i=1}^{n} u_i \frac{\partial u_l^2}{\partial x_i}\, dx = 0$$

by partial integration; observe that $\nabla\cdot u = \sum_{i=1}^{n}\frac{\partial u_i}{\partial x_i} = 0$. Although it is not allowed to insert u as a testing function in (IV.1.2), one can do so for the approximating solutions in the Galerkin approximation. Thus the Galerkin approximation just singles out weak solutions with energy inequality. These play a distinct rôle amongst the weak solutions in general because some sort of uniqueness holds for them as will be seen later on.

§ 2. Some Additional Regularity Properties for Weak Solutions in General

In this paragraph we want to connect Solonnikov's theorem III.1.1 (and its analogue for any dimension $n \geq 2$) and the weak solutions as they have been defined in definition IV.1.1. We start with an estimate for the nonlinearity.

<u>Proposition IV.2.1:</u> <u>Let</u> $T > 0$, <u>let</u> $u \in L^{\infty}((0,T),(L^2(\Omega))^n) \cap$ $\cap\, L^2((0,T),(\overset{\circ}{H}{}^{1,2}(\Omega))^n)$. <u>Then</u>

$$u\cdot\nabla u \in L^{\frac{n+2}{n+1}}((0,T),(L^{\frac{n+2}{n+1}}(\Omega))^n).$$

<u>Proof:</u> Let $\rho = 1 - \frac{2}{2+\frac{n}{4}}$. Then we have

$$|u\cdot\nabla u| \leq c(|\nabla u|^{2-\rho} + |u|^{(2-\rho)/(1-\rho)}),$$

$$\|u\|^{q'}_{(L^{q'}(\Omega))^n} \leq c\|u\|^{aq'}_{(\overset{\circ}{H}{}^{1,2}(\Omega))^n}\|u\|^{(1-a)q'}_{(L^2(\Omega))^{n'}}$$

$$\|\nabla u\|^{q''}_{(L^{q''}(\Omega))^{n^2}} \leq \|u\|^2_{(H^{1,2}(\Omega))^{n'}}$$

where $aq' = 2$, $q' = \frac{n+2}{n+1}\cdot\frac{2-\rho}{1-\rho}$, $q'' = \frac{n+2}{n+1}(2-\rho) = 2$ and where we have used the Gagliardo-Nirenberg inequalities in 0.2. □

For later considerations we give a second estimate for the nonlinear term $u\cdot\nabla u$, namely

Proposition IV.2.2: Let $p>1$, $u \in (H^{2,p}(\Omega))^n \cap (L^n(\Omega))^n$. Then

$$\|u\cdot\nabla u\|_{(L^p(\Omega))^n} \le c(\|u\|^3_{(L^{3p}(\Omega))^n} + \|\nabla u\|^{3/2}_{(L^{3p/2}(\Omega))^{n^2}}),$$

$$\le c\|u\|_{(H^{2,p}(\Omega))^n}(\|u\|^2_{(L^n(\Omega))^n} + \|u\|^{\frac{1}{2}}_{(L^n(\Omega))^n}).$$

Proof: First we have

$$|u\cdot\nabla u| \le c(|u|^3 + |\nabla u|^{3/2}).$$

Now we use again the Gagliardo-Nirenberg inequalities in 0.2. This yields

$$\|u\|^3_{(L^{3p}(\Omega))^n} \le c\|u\|_{(H^{2,p}(\Omega))^n}\|u\|^2_{(L^n(\Omega))^n},$$

$$\|\nabla u\|^{3/2}_{(L^{3p/2}(\Omega))^n} \le c\|u\|_{(H^{2,p}(\Omega))^n}\|u\|^{1/2}_{(L^n(\Omega))^n}.$$

The lemma is proved. □

According the theorem III.1.1 we can construct a solution (w,π) for

$$w' - \nu\Delta w + \nabla\pi = f - u\cdot\nabla u$$

if u is a given weak solution for the Navier-Stokes equation. It is our aim to identify w with the weak solution under consideration.

Definition IV.1.1: Let $p>1$, $T>0$, $\varphi \in (H_p(\Omega))^n$. An element $w \in L^p((0,T),(\overset{\circ}{H}{}^{1,p}(\Omega))^n)$ is called a weak solution with exponent p over $(0,T)\times\Omega$ of the equation

$$w' - \nu\Delta w + \nabla\tilde{\pi} = 0,$$
$$\nabla\cdot w = 0$$

with boundary values $w(0)=\varphi$, $w|\partial\Omega = 0$ *if and only if*

$$\text{(IV.2.1)} \quad -\int_0^T (w,\psi')\,dt + \nu\int_0^T (\nabla w,\nabla\psi)\,dt = (\varphi,\psi(0))$$

for all testing vectors ψ *with*

$\psi \in L^q((0,T),(\overset{\circ}{H}{}^{1,q}(\Omega))^n)$,

$\psi' \in L^q((0,T),(L^q(\Omega))^n)$,

$\nabla\cdot\psi = 0$ a.e. *in* $(0,T)$, $\psi(T) = 0$.

Here $\frac{1}{p}+\frac{1}{q} = 1$.

We first prove that any w with (IV.2.1) and $\varphi = 0$ vanishes identically and then make some remarks on definition IV.1.1.

Theorem IV.2.1: Let $p>1$, *let* $p \neq 3$, *let* $\varphi = 0$, $T>0$. *Let* w *be a weak solution as in definition* IV.1.1. *Then*

$w \equiv 0$.

Proof: According to theorem III.1.1 there is one and only one $\tilde\psi$ with

$\tilde\psi \in L^q((0,T),(H^{2,q}(\Omega))^n \cap (\overset{\circ}{H}{}^{1,q}(\Omega))^n)$,

$\tilde\psi' \in L^q((0,T),(L^q(\Omega))^n)$,

$\tilde\psi \in C^0([0,T],W^{2-2/q,q}(\Omega))$,

$\nabla\cdot\tilde\psi = 0$ a.e. on $(0,T)$

and one $\tilde\pi$ with

$\nabla\tilde\pi \in L^q((0,T),(L^q(\Omega))^n)$

such that

$$\tilde{\psi}' - \nu\Delta\tilde{\psi} + \nabla\tilde{\pi} = f(w(T-t)),$$
$$\nabla\cdot\tilde{\psi} = 0,$$
$$\tilde{\psi}(0) = 0.$$

Here q again is the exponent being dual to p; moreover we set

$$f_l(w(t)) = \begin{cases} |w(t)|^{p-1} \dfrac{w^l(t)}{|w(t)|}, & w(t) \neq 0, \\ 0 \text{ otherwise,} \end{cases}$$

$1 \leq l \leq n$. Therefore $f_l(w(.)) \in L^q((0,T),(L^q(\Omega))^n)$ and we get $\tilde{\psi},\tilde{\pi}$ as described just before. Now have to change $\tilde{\psi},\tilde{\pi}$ a little. We set

$$\tilde{\tilde{\psi}}(t) = \tilde{\psi}(T-t),$$
$$\tilde{\tilde{\pi}}(t) = \tilde{\pi}(T-t).$$

Then we have still

$$\tilde{\tilde{\psi}}'(t) = -\tilde{\psi}'(T-t), \; \tilde{\tilde{\psi}}' \in L^q((0,T),(L^q(\Omega))^n),$$
$$-\nu\Delta\tilde{\tilde{\psi}}(t) = -\nu\Delta\tilde{\psi}(T-t), \; \tilde{\tilde{\psi}} \in L^q((0,T),(H^{2,q}(\Omega))^n),$$
$$\nabla\cdot\tilde{\tilde{\psi}}(t) = \nabla\cdot\tilde{\psi}(T-t) = 0,$$
$$\nabla\tilde{\tilde{\pi}}(t) = \nabla\tilde{\pi}(T-t), \; \nabla\tilde{\tilde{\pi}} \in L^q((0,T),(L^q(\Omega))^n),$$
$$\tilde{\tilde{\psi}}(T) = 0,$$
$$-\tilde{\tilde{\psi}}' - \nu\Delta\tilde{\tilde{\psi}} + \nabla\tilde{\tilde{\psi}} = f(w(t)).$$

In particular $\tilde{\tilde{\psi}}$ is admissible as a testing vector in (IV.2.1). This gives

$$0 = \int_0^T (w,-\tilde{\tilde{\psi}}')\,dt + \nu\int_0^T (w,-\nu\Delta\tilde{\tilde{\psi}})\,dt$$
$$= \int_0^T (w,f(w)-\nabla\tilde{\tilde{\pi}})\,dt = \int_0^T (w,f(w))\,dt = \int_0^T \|w(t)\|^p_{(L^p(\Omega))^n}\,dt.$$

Thus it follows $w \equiv 0$. □

As for the definition IV.1.1 it first follows for a weak solution w that

$$w' \in L^p((0,T), H_*^{-1,p}(\Omega)).$$

Thus $w \in C^0([0,T], H_*^{-1,p}(\Omega))$ and the statement $w(0) = \varphi$ makes sense. As for the construction of $\tilde{\pi}$ this could be done in a way analogous to that we have used in connection with the definition of a weak solution of the Navier-Stokes equation.

For any weak solution of the Navier-Stokes equation as in definition IV.1.1 we now get the following regularity theorem:

<u>Theorem IV.2.2:</u> <u>Let</u> $T > 0$, <u>let</u>

$$\varphi \in H_2(\Omega) \cap (W^{2-\frac{2(n+1)}{n+2}, \frac{n+2}{n+1}}(\Omega)),$$

$$f \in L^2((0,T), (H^{-1,2}(\Omega))^n) \cap L^{\frac{n+2}{n+1}}((0,T), (L^{\frac{n+2}{n+1}}(\Omega))^n).$$

<u>Let</u> u <u>be a weak solution of the Navier-Stokes equations over</u> $(0,T) \times \Omega$ <u>with</u> $u(0) = \varphi$, $u|\partial\Omega = 0$ <u>in the sense of definition</u> IV.1.1. <u>Then</u>

$$u' \in L^{\frac{n+2}{n+1}}((0,T), (L^{\frac{n+2}{n+1}}(\Omega))^n),$$

$$u \in L^{\frac{n+2}{n+1}}((0,T), (H^{2,\frac{n+2}{n+1}}(\Omega))^n) \cap L^{\frac{n+2}{n+1}}((0,T), (\overset{\circ}{H}{}^{1,\frac{n+2}{n+1}}(\Omega))^n),$$

$$u \in C^0([0,T], (W^{2-\frac{2(n+1)}{n+2}, \frac{n+2}{n+1}}(\Omega))^n),$$

<u>there exists a pressure</u> π <u>such that</u>

$$\pi, \nabla\pi \in L^{\frac{n+2}{n+1}}((0,T), L^{\frac{n+2}{n+1}}(\Omega)),$$

<u>and we have</u>

$$u' - \nu\Delta u + u\cdot\nabla u + \nabla\pi = f$$
$$\nabla\cdot u = 0$$

<u>a.e. on</u> (0,T). π <u>is determined up to a function of</u> t <u>being in</u> $L^{\frac{n+2}{n+1}}((0,T),L^{\frac{n+2}{n+1}}(\Omega))$, <u>i.e. in</u> $L^{\frac{n+2}{n+1}}((0,T))$.

§ 3. On the validity of the Energy Inequality and on the Regularity of the Expression $u'+\nabla\pi$

In this chapter we want to pose various conditions on a weak solution of the Navier-Stokes equations under which the energy inequality holds. Let φ,f and any solution u of the Navier-Stokes equations to be considered be real valued from now on in this chapter. We start with a condition given by Serrin [Ser2].

<u>Theorem IV.3.1:</u> <u>Let</u> u <u>be a weak solution of the Navier-Stokes equations as in definition</u> IV.1.1. <u>Let additionally</u> u <u>be from</u>

$$L^s((0,T),(L^r(\Omega))^n)$$

<u>for some</u> s,r <u>with</u>

(IV.3.1) $\frac{n}{r}+\frac{2}{s} = 1,\ +\infty \geq r \geq n,\ +\infty \geq s \geq 2.$

<u>Then we have</u>

$$\text{(IV.3.2)}\quad \|u(t)\|^2_{H_2(\Omega)} + 2\nu \int_s^t \|\nabla u(\sigma)\|^2_{(L^2(\Omega))^{n^2}}\, d\sigma =$$

$$= \|u(s)\|^2_{H_2(\Omega)} + 2 \int_s^t (f(u(\sigma)),u(\sigma))\, d\sigma$$

<u>for all</u> $s,t,\ 0 \leq s \leq t \leq T$. <u>In particular</u> $\|u(.)\|_{H_2}$ <u>is absolutely continuous and</u>

$$u \in C^o([0,T],(L^2(\Omega))^n).$$

<u>Proof:</u> We estimate the nonlinearity. For a testing function ψ as in definition IV.1.1 we have

$$|(u(t)\cdot u(t),\nabla\psi(t))|$$

$$\leq \|\nabla\psi(t)\|_{(L^2(\Omega))^{n^2}}\|u(t)\|_{(L^r(\Omega))^n}\|u(t)\|_{(L^{\tilde{s}}(\Omega))^n},$$

where $\frac{1}{r}+\frac{1}{\tilde{s}}=\frac{1}{2}$. Let $r>n$, $s>2$. The Gagliardo-Nirenberg inequality of our auxiliary propositions gives

$$\|u(t)\|_{(L^{\tilde{s}}(\Omega))^n} \leq c\|\nabla u(t)\|^a_{(L^2(\Omega))^{n^2}}\|u(t)\|^{1-a}_{(L^2(\Omega))^n}$$

with $\frac{1}{\tilde{s}} = a(\frac{1}{2}-\frac{1}{n}) + (1-a)\frac{1}{2} = \frac{1}{2}-\frac{a}{n}$. Thus $\frac{a}{n} = \frac{1}{2}-\frac{1}{\tilde{s}}$, $a = \frac{n}{2}-\frac{n}{\tilde{s}}$ and, because $\frac{1}{\tilde{s}} = \frac{1}{2}-\frac{1}{r}$, $a=\frac{n}{r}$; (IV.3.1) yields $1-a = \frac{2}{s}$ and we arrive at

$$\|u(t)\|_{(L^{\tilde{s}}(\Omega))^n} \leq c\|\nabla u(t)\|^{\frac{n}{r}}_{(L^2(\Omega))^{n^2}}\|u(t)\|^{\frac{2}{s}}_{(L^2(\Omega))^n}.$$

This gives

$$|(u(t)\cdot u(t),\nabla\psi(t))| \leq \|\nabla\psi(t)\|_{(L^2(\Omega))^{n^2}} \|u(t)\|_{(L^r(\Omega))^n}\cdot$$

$$\cdot\|\nabla u(t)\|^{\frac{n}{r}}_{(L^2(\Omega))^{n^2}} \|u(t)\|^{\frac{2}{s}}_{(L^2(\Omega))^{n'}}$$

$$\text{(IV.3.3)} \quad \int_0^T |(u(t)\cdot u(t),\nabla\psi(t))|\, dt \leq \left(\int_0^T \|\nabla\psi(t)\|^2_{(L^2(\Omega))^{n^2}}\, dt\right)^{\frac{1}{2}}$$

$$\cdot\left(\int_0^T \|u(t)\|^s_{(L^r(\Omega))^n}\, dt\right)^{\frac{1}{s}}\cdot$$

$$\cdot\left(\int_0^T \|\nabla u(t)\|^2_{(L^2(\Omega))^{n^2}} dt\right)^{\frac{s-2}{2s}} \cdot \operatorname*{ess\,sup}_{0<t<T} \|u(t)\|^{\frac{2}{s}}_{(L^2(\Omega))^n}.$$

Now let $\tilde{\psi} \in L^2((0,T),(\overset{o}{H}{}^{1,2}(\Omega))^n)$, $\psi' \in L^2((0,T),(L^2(\Omega))^n)$, $\nabla\cdot\psi(t) = 0$, $\tilde{\psi}(T) = 0$. As was pointed out in 0.2 any such $\tilde{\psi}$ can be approximated in the norm

$$\left(\int_0^T \|\tilde{\psi}'(t)\|^2_{(L^2(\Omega))^n}\, dt + \int_0^T \|\nabla\tilde{\psi}(t)\|^2_{(L^2(\Omega))^{n^2}}\, dt\right)^{\frac{1}{2}}$$

by testing functions as introduced in definition IV.1.1. (IV.3.3) then simply means that

$$u' \in L^2((0,T),H_*^{-1,2}(\Omega))$$

and we have

$$u' - \nu\Delta u + u\cdot\nabla u = f \text{ a.e. on } (0,T)$$

as an equation in $L^2((0,T),(H^{-1,2}(\Omega))^n)$. Applying both sides on $u \in L^2((0,T),\overset{o}{H}{}^{1,2}(\Omega))$ and taking into consideration that for almost all $t \in (0,T)$

$$(u(t)\cdot\nabla u(t),u(t)) = 0$$

(the partial integration can be carried through as at the end of IV.1 since the chain rule holds for $\frac{\partial u_1^2}{\partial x_i}$ as the reader may verify by himself; also the integrals are well defined as the preceding calculations show), we get the result requested by using (0.2.6) of our auxiliary propositions.

We still have to deal with the cases $r = n$ or $s = 2$. If $r = n$ then necessarily $s = +\infty$, and we have the estimate

$$\text{(IV.3.4)}\quad \int_0^T |(u(t)\cdot u(t),\nabla\psi(t))|\, dt \leq$$

$$\leq \int_0^T \|u(t)\|_{(L^n(\Omega))^n}\|u(t)\|_{(L^{\frac{2n}{n-2}}(\Omega))^n}\|\nabla\psi(t)\|_{(L^2(\Omega))^{n^2}}\, dt,$$

$$\leq c \operatorname*{ess\,sup}_{0<t<T} \|u(t)\|_{(L^n(\Omega))^n}\left(\int_0^T \|\nabla u(t)\|^2_{(L^2(\Omega))^{n^2}}\, dt\right)^{\frac{1}{2}}\cdot$$

$$\cdot\left(\int_0^T \|\nabla\psi(t)\|^2_{(L^2(\Omega))^{n^2}}\, dt\right)^{\frac{1}{2}}$$

by Sobolev's imbedding theorem. Thus the same conclusion holds as before. If $s = 2$ then $r = +\infty$ and we get

$$\text{(IV.3.5)}\quad \int_0^T |(u(t)\ u(t),\nabla\psi(t))| \leq c\left(\int_0^T \|u(t)\|^2_{(L\ (\Omega))^n}\, dt\right)^{\frac{1}{2}}$$

$$\operatorname*{ess\,sup}_{0<t<T} \|u(t)\|_{(L^2(\Omega))^n}\cdot\left(\int_0^T \|\nabla\psi(t)\|^2_{(L^2(\Omega))^{n^2}}\, dt\right)^{\frac{1}{2}};$$

thus our theorem is also proved in this case. □

This theorem is essentially due to Serrin [Ser2, theorem 5]. Serrin also has introduced the "critical quantity" $\frac{n}{r}+\frac{2}{s}$ which will play an important rôle later on. For a different proof see [SOW 1, §3]. We remark that the estimates employed for the proof of theorem IV.3.1 yield a slight generalization, namely we get

$$\text{(IV.3.6)}\quad \int_0^T |(u(t)\cdot v(t),\nabla\psi(t))|\,dt$$

$$\leq c\Big(\int_0^T \|u(t)\|^s_{(L^r(\Omega))^n}\cdot\|v(t)\|^2_{(L^2(\Omega))^n}\Big)^{\frac{1}{s}}\cdot$$

$$\cdot\Big(\int_0^T \|\nabla v(t)\|^{\frac{n}{r}\frac{s}{s-1}}_{(L^2(\Omega))^{n^2}}\|\nabla\psi(t)\|^{\frac{s}{s-1}}_{(L^2(\Omega))^{n^2}}\,dt\Big)^{\frac{s-1}{s}}$$

if $\frac{n}{r}+\frac{2}{s}=1$, $+\infty\geq r\geq n$, $+\infty\geq s\geq 2$, $r>n$, $s>2$. u is as in theorem IV.3.1 and v is in $L^\infty((0,T),(L^2(\Omega))^n)\cap L^2((0,T),(\overset{\circ}{H}{}^{1,2}(\Omega))^n)$. For $r=n$ we have then also

$$\text{(IV.3.7)}\quad \int_0^T |(u(t)\cdot v(t),\nabla\psi(t))|\,dt \leq \operatorname*{ess\,sup}_{0<t<T}\|u(t)\|_{(L^n(\Omega))^n}$$

$$\cdot\Big(\int_0^T \|\nabla v(t)\|_{(L^2(\Omega))^{n^2}}\|\nabla\psi(t)\|_{(L^2(\Omega))^{n^2}}\,dt\Big),$$

and for $s=2$ we get

$$\text{(IV.3.8)}\quad \int_0^T |(u(t)\cdot v(t),\nabla\psi(t))|\,dt \leq$$

$$\leq c\int_0^T \|u(t)\|_{(L^\infty(\Omega))^n}\|v(t)\|_{(L^2(\Omega))^n}\|\nabla\psi(t)\|_{(L^2(\Omega))^{n^2}}\,dt.$$

Now we want to study the properties of the pressure belonging to a weak solution of the Navier-Stokes equation as introduced in definition IV.1.1. Theorem IV.2.2 already shows that $\pi\in L^{\frac{n+2}{n+1}}((0,T),L^{\frac{n+2}{n+1}}(\Omega))$ but a still stronger result is true.

Theorem IV.3.2: Let u be a weak solution of the Navier-Stokes equations over $(0,T)\times\Omega$ with boundary values $u(0)=\varphi$, $u|\partial\Omega=0$ in the sense of definition IV.1.1. Additionally to the assumptions $f\in L^2((0,T),(H^{-1,2}(\Omega))^n)$, $\varphi\in H_2(\Omega)$, we assume that

$$f\in L^{\frac{n+2}{n+1}}((0,T),(L^{\frac{n+2}{n+1}}(\Omega))^{\frac{n+2}{n+1}}),\quad \varphi\in W^{2-2/\frac{n+2}{n+1},\frac{n+2}{n+1}}(\Omega).$$

We take π from theorem IV.2.2, i.e. $\pi\in L^{\frac{n+2}{n+1}}((0,T),L^{\frac{n+2}{n+1}}(\Omega))$; then $u'+\nabla\pi\in L^{\frac{n+2}{n}}((0,T),(H^{-1,\frac{n+2}{n}}(\Omega))^n)$ and $\pi\in L^{\frac{n+2}{n}}((0,T),L^{\frac{n+2}{n}}(\Omega))$ if $u'\in L^{\frac{n+2}{n}}((0,T),(H^{-1,\frac{n+2}{n}}(\Omega))^n)$. Moreover

$$\text{(IV.3.9)}\quad \pi\in L^{\frac{n+2}{n+1}}((0,T),L^{\frac{n(n+2)}{n^2-2}}(\Omega)),$$

$$\text{(IV.3.10)}\quad u\in L^{\frac{2(n+2)}{n}}((0,T),L^{\frac{2(n+2)}{n}}(\Omega)),$$

$$u'\in L^{\frac{n+2}{n+1}}((0,T),(H^{-1,\frac{n+2}{n}}(\Omega))^n).$$

Proof: For the nonlinearity we obtain (use the Gagliardo-Nirenberg inequality from 0.2)

$$\|u\cdot u\|_{(L^p(\Omega))^{n^2}}\le c\|u\|^2_{(L^{2p}(\Omega))^n}$$

$$\le c\|\nabla u\|^{2a}_{(L^2(\Omega))^{n^2}}\|u\|^{2(1-a)}_{(L^2(\Omega))^n}$$

with $p=\frac{n+2}{n}$, $\frac{1}{2p}=\frac{n}{2(n+2)}=a(\frac{1}{2}-\frac{1}{n})+(1-a)\frac{1}{2}=\frac{1}{2}-a\frac{1}{n}$ if $a=\frac{n}{n+2}$. This means that

$$\int_0^T |(u(t)\cdot u(t),\nabla\psi(t))|\,dt\le$$

$$\le\Big(\int_0^T\|\nabla\psi(t)\|^{\frac{n+2}{2}}_{(L^{\frac{n+2}{2}}(\Omega))^{n^2}}dt\Big)^{\frac{2}{n+2}}\cdot\Big(\int_0^T\|\nabla u(t)\|^2_{(L^2(\Omega))^{n^2}}dt\Big)^{\frac{n}{n+2}}\cdot$$

$$\cdot\operatorname*{ess\,sup}_{0<t<T}\|u(t)\|^{4/(n+2)}_{(L^2(\Omega))^n},$$

$\psi \in L^{\frac{n+2}{2}}((0,T),(\overset{\circ}{H}{}^{1,\frac{n+2}{2}}(\Omega))^n)$. Thus we get

$$\text{(IV.3.11)}\quad u\cdot\nabla u \in L^{\frac{n+2}{n}}((0,T),(H^{-1,\frac{n+2}{n}}(\Omega))^n).$$

Obviously

$$\text{(IV.3.12)}\quad -\nu\Delta u \in L^{\frac{n+2}{n}}((0,T),(H^{-1,\frac{n+2}{n}}(\Omega))^n),$$

$$\text{(IV.3.13)}\quad u' \in L^{\frac{n+2}{n}}((0,T),H_*^{-1,\frac{n+2}{n}}(\Omega)),$$

$$\text{(IV.3.14)}\quad f \in L^{\frac{n+2}{n}}((0,T),(H^{-1,\frac{n+2}{n}}(\Omega))^n).$$

Because of our additional assumptions on φ, f we get (according to theorem IV.2.2) that

$$u' - \nu\Delta u + u\cdot\nabla u + \nabla\pi = f$$

a.e. on $(0,T)$. Thus $u'+\nabla\pi \in L^{\frac{n+2}{n}}((0,T),(H^{-1,\frac{n+2}{n}}(\Omega))^n)$,

$$\text{(IV.3.15)}\quad \nabla\pi \in L^{\frac{n+2}{n}}((0,T),(H^{-1,\frac{n+2}{n}}(\Omega))^n)$$

if it could be shown that $u' \in L^{\frac{n+2}{n}}((0,T),(H^{-1,\frac{n+2}{n}}(\Omega))^n)$. Theorem III.2.6 does not furnish this inclusion, although we know already the inclusion (IV.3.13), and although

$$|(u'(t),\psi)| \le \|u'(t)\|_{(L^{\frac{n+2}{n+1}}(\Omega))^n} \|\psi\|_{(L^{n+2}(\Omega))^n}$$

$$\le c\|u'(t)\|_{(L^{\frac{n+2}{n+1}}(\Omega))^n} \|\nabla\psi\|_{(L^{\frac{n+2}{2}}(\Omega))^{n^2}}$$

by Sobolev and consequently $u' \in L^{\frac{n+2}{n+1}}((0,T),(H^{-1,\frac{n+2}{n}}(\Omega))^n)$. The dual of the space $L^p_M(\Omega)$ is $L^q(\Omega)/\mathbb{C}$, $\frac{1}{p}+\frac{1}{q}=1$, $1<p,q<+\infty$. $L^2(\Omega)/\mathbb{C}$ is isometrically isomorphic to $L^2_M(\Omega)$ (for the definition of

$L^p_M(\Omega)$ cf. 0.2, p. XVI). If the remainder classes of $L^q(\Omega)/\mathbb{C}$ are denoted by $\{f+\tilde{c}\}$, then

(IV.3.16) $$\|\{f+\tilde{c}\}\|_{L^q(\Omega)/\mathbb{C}} = \inf_{\tilde{c}\in\mathbb{C}} \|f+\tilde{c}\|_{L^q(\Omega)},$$

and we have in the general case

(IV.3.17) $$\frac{1}{c}\|\{f+\tilde{c}\}\|_{L^p(\Omega)/\mathbb{C}} \le \|f\|_{L^p_M(\Omega)} \le c\|\{f+\tilde{c}\}\|_{L^p(\Omega)/\mathbb{C}},$$

$f\in L^p_M(\Omega)$; c is a positive constant depending on p. This is seen by showing that the mapping: $f\mapsto\{f+\tilde{c}\}$ is a bounded linear operator and a bijection. We also consider the closed subspace $N = \{g\,|\,g\in(L^q(\Omega))^n,\ \nabla\cdot B_q^{-1/2}g=0\}$ of $(L^q(\Omega))^n$ and $(L^q(\Omega))^n/N$. The dual space of $(L^q(\Omega))^n/N$ consists of the elements $f\in(L^p(\Omega))^n$ vanishing on N. If $f\in((L^q(\Omega))^n/N)^*$ then

(IV.3.18) $$\sup_{\substack{u\neq 0,\\ u\in(L^q(\Omega))^n/N}} \frac{|f(u)|}{\inf_{v\in N}\|u+v\|_{(L^q(\Omega))^n}}$$

$$= \sup_{\substack{u\neq 0,\\ u\in(L^q(\Omega))^n/N}} \sup_{v\in N} \frac{|f(u+v)|}{\|u+v\|_{(L^q(\Omega))^n}} \le \|f\|_{(L^p(\Omega))^n}.$$

We define two operators H and $\tilde{H}$ by setting

$$H: L^p_M(\Omega)\supset H^{1,p}_M(\Omega)=D(H)\to(L^p(\Omega))^n,\quad f\mapsto B_p^{-1/2}\nabla f,$$

$$\tilde{H}: (L^q(\Omega))^n/N\to L^q(\Omega)/\mathbb{C},\quad g+N\mapsto\{\nabla\cdot B_q^{-1/2}g+\tilde{c}\}.$$

From what was said on p. 96 on $B_q^{1/2}$ we get that $\tilde{H}$ is bounded. According to the result of Bogovskij-Erig (cf. 0.2, p. XVI) $\tilde{H}$ is onto. $\tilde{H}$ is also unique: Assume that $\{\nabla\cdot B_q^{-1/2}g+\tilde{c}\}=0$. Then $\nabla\cdot B_q^{-1/2}g=0$ since $\int_\Omega \nabla\cdot B_q^{-1/2}g\,dx=0$. Thus H has a bounded inverse, and consequently

$$\tilde{H}^*: L^p_M(\Omega)\to((L^q(\Omega))^n/N)^*$$

is bounded and has a bounded inverse $\tilde{H}^{*-1}=(\tilde{H}^{-1})^*$. For $f\in D(H)$ we obtain $Hf=\tilde{H}^*f$ and with what was said on p. 96 on $B_p^{1/2}$ and (IV.3.18) we get $\|\nabla f\|_{(H^{-1,p}(\Omega))^n}\ge c\|B_p^{-1/2}\nabla f\|_{(L^p(\Omega))^n}\ge$

$c\|\tilde{H}^* f\|_{((L^q(\Omega))^n/N^*)} \geq c\|f\|_{L^p_M(\Omega)}$. For $p = (n+2)/n$ we thus arrive at

$$\|\pi(t) - \frac{\int_\Omega \pi(t)\,dx}{\text{mes }\Omega}\|_{L^{\frac{n+2}{n}}(\Omega)} \leq c\|\nabla\pi(t)\|_{(H^{-1,\frac{n+2}{n}}(\Omega))^n}.$$

As for the proof of the property (IV.3.10) of u we get as in the beginning of the proof that

$$\|u(t)\|^{2p}_{(L^{2p}(\Omega))^n} \leq c\|\nabla u(t)\|^{2ap}_{(L^2(\Omega))^n}\|u\|^{2(1-a)p}_{(L^2(\Omega))^n}$$

with $p = \frac{n+2}{n}$, $a = \frac{1}{p}$. (IV.3.9) is simply a consequence of theorem IV.2.2 and Sobolev's imbedding theorem. Our theorem is proved. □

Although Ω is bounded throughout we want to mention that in the case $\Omega = \mathbb{R}^n$ (the weak solution is, of course, defined analogously) there is a very simple proof in order to show that

$$\pi \in L^{\frac{n+2}{n}}((0,T), L^{\frac{n+2}{n}}_{loc}(\mathbb{R}^n)) + L^2((0,T), L^2_{loc}(\mathbb{R}^n)).$$

This will be outlined in the comments to this chapter.

It would be desirable to have a criterion for the validity of the energy inequality weaker than (IV.3.1); by that we mean that $\frac{n}{r}+\frac{2}{s} > 1$. We propose the assumption

$$u \cdot u \in L^2((0,T), (L^2(\Omega))^{n^2})$$

or equivalently

$$u \in L^4((0,T), (L^4(\Omega))^n).$$

In this case $\frac{n}{r}+\frac{2}{s} = \frac{n}{4}+\frac{2}{4} = \frac{n}{4}+\frac{1}{2} = \frac{5}{4}$ for $n = 3$ ($\geq \frac{3}{2}$ for $n \geq 4$).

<u>Theorem IV.3.3:</u> <u>Let</u> u <u>be a weak solution of the Navier-Stokes equations over</u> $(0,T)\times\Omega$ <u>with</u> $u(0) = \varphi$, $u|\partial\Omega = 0$ <u>in the sense of definition</u> IV.1.1. <u>Let</u>

$$u\cdot u \in L^2((0,T),(L^2(\Omega))^{n^2})$$

<u>or equivalently let</u>

$$u \in L^4((0,T),(L^4(\Omega))^n).$$

<u>Then we have</u>

$$\|u(t)\|^2_{H_2(\Omega)} + 2\nu \int_s^t \|\nabla u(\sigma)\|^2_{(L^2(\Omega))^{n^2}} d\sigma =$$

$$= \|u(s)\|^2_{H_2(\Omega)} + 2\int_s^t (f(\sigma),u(\sigma))\, d\sigma$$

<u>for all</u> $s,t,\ 0 < s \leq t \leq T$. <u>In particular</u> $\|u(.)\|_{H_2(\Omega)}$ <u>is absolutely continuous and consequently</u>

$$u \in C^0([0,T],H_2(\Omega)).$$

<u>If we additionally assume that</u> $n = 3$,

$$\varphi \in (W^{2-2/(5/4),5/4}(\Omega))^n = (W^{2/5,5/4}(\Omega))^n,$$

$$f \in L^{5/4}((0,T),(L^{5/4}(\Omega))^n)$$

<u>then</u> $u' \in L^{5/4}((0,T),(H^{-1,2}(\Omega))^n)$. <u>Moreover</u>

$$u'+\nabla\pi \in L^2((0,T),(H^{-1,2}(\Omega))^n)$$

<u>and</u> $\pi \in L^2((0,T),L^2(\Omega))$ <u>if</u> $u' \in L^2((0,T),(H^{-1,2}(\Omega))^n)$.

<u>Proof:</u> We start with an estimate of the nonlinearity. We have

$$|(u(t)\cdot u(t),\nabla\psi(t))| \leq c\|u(t)\cdot u(t)\|_{(L^2(\Omega))^{n^2}} \cdot \|\nabla\psi(t)\|_{(L^2(\Omega))^{n^2}},$$

$$\int_0^T |(u(t)\cdot u(t),\nabla\psi(t))|\, dt \leq c\left(\int_0^T \|u(t)\cdot u(t)\|^2_{(L^2(\Omega))^{n^2}} dt\right)^{\frac{1}{2}}$$

$$\left(\int_0^T \|\nabla\psi(t)\|^2_{(L^2(\Omega))^{n^2}} dt\right)^{\frac{1}{2}},$$

$\psi \in L^2((0,T),\overset{\circ}{H}{}^{1,2}(\Omega))^n)$. Thus

$$u\cdot\nabla u(t) \in L^2((0,T),(H^{-1,2}(\Omega))^n).$$

The remaining part of the proof of the energy equality can be carried through as in the proof of theorem IV.3.1. As for the regularity property of the pressure we can argue as in the second part of the proof of theorem IV.3.2. First the equation

$$u' - \nu\Delta u + u\cdot\nabla u + \nabla\pi = f,$$

which holds in $L^{5/4}((0,T),(L^{5/4}(\Omega))^n)$, shows that

$$u'+\nabla\pi \in L^2((0,T),(H^{-1,2}(\Omega))^n).$$

Then because of $u' \in L^{5/4}((0,T),(L^{5/4}(\Omega))^n)$ (observe that $\frac{5}{4} = \frac{n+2}{n+1}$ for $n = 3$) we can show as in the proof of theorem IV.3.2 that

$$u' \in L^{5/4}((0,T),(H^{-1,2}(\Omega))^n).$$

Let

$$\nabla\pi \in L^2((0,T),(H^{-1,2}(\Omega))^n).$$

Now we can use a result of Nečas [Tem, p. 15] which assures that

$$\pi \in L^2((0,T),L^2(\Omega))$$

but we can also prove this inclusion by using the corresponding argument in the proof of theorem IV.3.2; it is even simpler here since we are in the Hilbert space case. □

<u>Remark:</u> The assumptions $\varphi \in (W^{2-2/\frac{n+2}{n+1}}(\Omega))^n$, $\varphi \in (W^{2/5,5/4}(\Omega))^n$ in theorems IV.3.2, IV.3.3 respectively have only been made not to overburden the proof. <u>In fact we can drop them</u>; we will sketch the necessary considerations: If u is a weak solution as in definition IV.1.1 then

$$u(t) \in (\overset{\circ}{H}{}^{1,2}(\Omega))^n,\ \nabla\cdot u(t) = 0 \text{ a.e. on } (0,T),$$

and consequently

$$u(t) \in (W^{2-2/\frac{n+2}{n+1},\frac{n+2}{n+1}}(\Omega))^n \cap H_{\frac{n+2}{n+1}}(\Omega)$$

a.e. on $(0,T)$. Thus considering u as a weak solution over $(\varepsilon,T)\times\Omega$ with boundary values $u(\varepsilon) \in (W^{2-2/\frac{n+2}{n+1},\frac{n+2}{n+1}}(\Omega))^n \cap H_{\frac{n+2}{n+1}}(\Omega) \cap H_2(\Omega)$, $u|\partial\Omega=0$ we can apply our former argument and get

$$u'+\nabla\pi \in L^{\frac{n+2}{n}}((\varepsilon,T),(H^{-1,\frac{n+2}{n}}(\Omega))^n),\ u'+\nabla\pi \in L^2((\varepsilon,T),(H^{-1,2}(\Omega))^n).$$

Let $u' \in L^{\frac{n+2}{n}}((0,T),(H^{-1,\frac{n+2}{n}}(\Omega))^n)$, $L^2((0,T),(H^{-1,2}(\Omega))^n)$ respectively. Since we have the estimate

$$\int_\varepsilon^T \|\pi(t) - \frac{1}{\text{mes }\Omega}\int_\Omega \pi(t)\,dx\|^{\frac{n+2}{n}}_{L^{\frac{n+2}{n}}(\Omega)}\,dt \le$$

$$\le c\int_\varepsilon^T \|\nabla\pi(t)\|^{\frac{n+2}{n}}_{(H^{-1,\frac{n+2}{n}})^n}\,dt \le c\Big(\int_\varepsilon^T \|u'(t)\|^{\frac{n+2}{n}}_{(H^{-1,\frac{n+2}{n}})^n}\,dt +$$

$$+ \int_\varepsilon^T \|\Delta u(t)\|^{\frac{n+2}{n}}_{(H^{-1,\frac{n+2}{n}})^n}\,dt + \int_\varepsilon^T \|u(t)\cdot\nabla u(t)\|^{\frac{n+2}{n}}_{(H^{-1,\frac{n+2}{n}})^n}\,dt\Big)$$

and a corresponding one for $\int_\varepsilon^T \|\pi(t) - \int_\Omega \pi(t)dx\|^2_{L^2(\Omega)}\,dt$ (the constants c do not depend on ε), we can let ε tend to 0 to get the result requested. Also the <u>additional assumption</u>

$$f \in L^{\frac{n+2}{n+1}}((0,T),(L^{\frac{n+2}{n+1}}(\Omega))^n)$$

can be weakened (concerning $u'+\nabla\pi$). It has been used to ensure that the equation

$$u' - \nu\Delta u + u\cdot\nabla u + \nabla\pi = f$$

holds in a suitable sense. But this conclusion also follows if

$$f \in L^{\frac{n+2}{n+1}}((T',T''),(L^{\frac{n+2}{n+1}}(\Omega))^n$$

on every interval (T',T''), $0 < T' < T'' < T$. □

§ 4. On the Uniqueness of Weak Solutions. The Connection between Weak Solutions and Local Strong Solutions

First we will prove the following generalization of Serrin's uniqueness result [Ser2, Theorem 6] (Recall that the data and the weak solutions under consideration are supposed to be real valued):

Theorem IV.4.1: Let u^1, u^2 be two weak solutions of the Navier-Stokes equations over $(0,T)\times\Omega$ as in definition VI.1.1 with boundary values $u^1(0) = \varphi^1$, $u^2(0) = \varphi^2$, $u^1|\partial\Omega = 0$, $u^2|\partial\Omega = 0$ and with inhomogeneous terms f^1, f^2. Let the energy inequalities

$$\text{(IV.4.1)} \quad \|u^i(t)\|^2_{H_2(\Omega)} + 2\nu \int_0^t \|\nabla u^i(\sigma)\|^2_{(L^2(\Omega))^{n^2}} d\sigma \le$$

$$\le \|\varphi^i\|^2_{H^2(\Omega)} + 2\int_0^t (f^i(\sigma), u(\sigma))\, d\sigma, \quad i = 1,2,$$

hold for all t, $0 \le t \le T$. One of the u^i, say u^1, fulfills the condition

$$\text{(IV.4.2)} \quad u^1 \in L^s((0,T),(L^r(\Omega))^n)$$

for some r,s *with* $n \leq r \leq +\infty$, $2 \leq s < \infty$, $\frac{2}{s}+\frac{n}{r}=1$. *Then the following estimates are valid:*

$$\|u^1(t)-u^2(t)\|^2_{H_2(\Omega)} + 2\nu \int_0^t \|\nabla u^1(\sigma)-\nabla u^2(\sigma)\|^2_{(L^2(\Omega))^{n^2}} d\sigma$$

$$\leq \|\varphi^1-\varphi^2\|^2_{H_2(\Omega)} + 2\int_0^t (f_1(\sigma)-f_2(\sigma),u^1(\sigma)-u^2(\sigma))\, d\sigma +$$

$$+ c\Big(\int_0^t \|u^1(\sigma)\|^{\frac{2r}{r-n}}_{(L^r(\Omega))^n}\|u^1(\sigma)-u^2(\sigma)\|^2_{(L^2(\Omega))^n} d\sigma\Big)^{\frac{r-n}{2r}} \cdot$$

$$\cdot \Big(\int_0^t \|\nabla u^1(\sigma)-\nabla u^2(\sigma)\|^2_{(L^2(\Omega))^{n^2}} d\sigma\Big)^{\frac{1}{2}+\frac{n}{2r}}$$

if $\frac{n}{r}+\frac{2}{s}=1$, $n<r<+\infty$, $2<s<+\infty$, *and consequently* $s=\frac{2r}{r-n}$,

$$\|u^1(t)-u^2(t)\|^2_{H_2(\Omega)} + 2\nu \int_0^t \|\nabla u^1(\sigma)-\nabla u^2(\sigma)\|^2_{(L^2(\Omega))^{n^2}} d\sigma$$

$$\leq \|\varphi^1-\varphi^2\|^2_{H_2(\Omega)} + 2\int_0^t (f_1(\sigma)-f_2(\sigma),u^1(\sigma)-u^2(\sigma))\, d\sigma +$$

$$+ \operatorname*{ess\,sup}_{0<\sigma<t} \|u^1(\sigma)\|_{(L^r(\Omega))^n} \int_0^t \|\nabla u^1(\sigma)-\nabla u^2(\sigma)\|^2_{(L^2(\Omega))^{n^2}} d\sigma$$

if $\frac{n}{r}+\frac{2}{s}=1$, $n=r$, $s=+\infty$,

$$\|u^1(t)-u^2(t)\|^2_{H_2(\Omega)} + 2\nu \int_0^t \|\nabla u^1(\sigma)-\nabla u^2(\sigma)\|^2_{(L^2(\Omega))^{n^2}} d\sigma$$

$$\leq \|\varphi_1-\varphi_2\|^2_{H_2(\Omega)} + 2\int_0^t (f_1(\sigma)-f_2(\sigma),u^1(\sigma)-u^2(\sigma))\, d\sigma +$$

$$+ c\int_0^t \|u^1(\sigma)\|_{(L^\infty(\Omega))^n}\|u^1(\sigma)-u^2(\sigma)\|_{(L^2(\Omega))^n}$$

$$\|\nabla u^1(\sigma)-\nabla u^2(\sigma)\|_{(L^2(\Omega))^{n^2}} d\sigma$$

if $\frac{n}{r}+\frac{2}{s}=1$, $r=+\infty$, $s=2$.

In particular we get in the first and in the third case: $u^1 = u^2$ if $\varphi^1 = \varphi^2$, $f^1 = f^2$. As for the second case we have $u^1 = u^2$ if $\varphi^1 = \varphi^2$, $f^1 = f^2$ and if

$$u^1 \in C^o([0,T], (L^n(\Omega))^n).$$

Proof: We consider u^1, u^2 as weak solutions over $(0,\tilde{T}) \times \Omega$, $0 < T \leq \tilde{T}$, rather than over $(0,T) \times \Omega$ (cf. theorem IV.1.1). The idea (see [So1]) is to approximate any weak solution by its Yosida approximations and then to use Serrin's idea of comparing two solutions. The proof has been given in [SOW1]. Let $\overline{n} = [\frac{n}{2}]+1$,

$$\tilde{J}_m = (I + \frac{1}{m} A_2)^{-[\frac{n}{2}]-1}, \quad m \in \mathbb{N},$$

where A_2 is the positive selfadjoint Stokes operator $-\nu P\Delta$ in $H_2(\Omega)$ with domain of definition $(H^{2,2}(\Omega))^n \cap (\overset{\circ}{H}{}^{1,2}(\Omega))^n \cap H_2(\Omega)$. From the regularity theorem (III.1.10) it follows: $D(A_2^{(\overline{n}+i)/2}) \subset (H^{\overline{n}+i,2}(\Omega))$

$$\delta \| u \|_{(H^{\overline{n}+i,2}(\Omega))^n} \leq \| A_2^{\frac{\overline{n}}{2}+\frac{i}{2}} u \| \leq \frac{1}{\delta} \| u \|_{(H^{\overline{n}+i,2}(\Omega))^n}, \quad i=0,1,$$

for some $\delta > 0$

and all $u \in D(A_2^{\frac{\overline{n}+i}{2}})$. Let u be any weak solution as in definition IV.1.1. Let ψ be a testing function with

$$\psi \in C^{o,1}([0,\tilde{T}], (L^2(\Omega))^n),$$
$$\psi \in C^o([0,\tilde{T}], (\overset{\circ}{H}{}^{1,2}(\Omega))^n),$$

$\nabla \cdot \psi = 0$ on $[0,\tilde{T}]$ and $\psi(\tilde{T}) = 0$. Then we consider $A_2^{-\overline{n}} \psi$; of course also $A^{-\overline{n}} \psi \in C^{o,1}([0,\tilde{T}], (L^2(\Omega))^n)$, $A_2^{-\overline{n}} \psi \in C^o([0,\tilde{T}], (\overset{\circ}{H}{}^{1,2}(\Omega))^n)$, but since by Sobolev

$$\| \nabla A_2^{-\overline{n}} \psi(t) \|_{(L^2(\Omega))^n} \leq \| \nabla A_2^{-\frac{\overline{n}}{2}} \psi(t) \|_{(C^o(\overline{\Omega}))^n},$$

$$\leq c \| A_2^{-\frac{\overline{n}}{2}} \psi(t) \|_{(C^1(\overline{\Omega}))^n} \leq c \| A_2^{\frac{1}{2}} \psi(t) \|_{H_2(\Omega)} \leq c \| \nabla \psi(t) \|_{H_2(\Omega)}$$

we see that $A_2^{-\bar{n}}\psi$ is an admissible testing function in the definition IV.1.1. Thus $\tilde{J}_m\psi$ is admissible, and we get

$$-\int_0^{\tilde{T}} (\tilde{J}_m u(t),\psi'(t))\,dt + \nu\int_0^{\tilde{T}} (\nabla u(t),\nabla\tilde{J}_m\psi(t))\,dt - \int_0^{\tilde{T}} (u(t)\cdot u(t),\nabla\tilde{J}_m\psi(t))\,dt =$$

$$= \int_0^{\tilde{T}} (f(t),\tilde{J}_m\psi(t))\,dt + (\varphi,\tilde{J}_m\psi(0)).$$

Now $\nu(\nabla u(t),\nabla\tilde{J}_m\psi(t)) = (u(t),A_2\tilde{J}_m\psi(t)) = (u(t),\tilde{J}_m A_2\psi(t)) = (A_2\tilde{J}_m u(t),\psi(t))$, $-(u(t)\cdot u(t),\nabla\tilde{J}_m\psi(t)) = \langle P_{\frac{n+2}{n+1}}(u(t)\cdot\nabla u(t)),\tilde{J}_m\psi(t)\rangle = (\tilde{J}_m P_{\frac{n+2}{n+1}}(u(t)\cdot\nabla u(t)),\psi(t))$; here $\langle .,.\rangle$ is the $L^{\frac{n+2}{n+1}}-L^{n+2}$ duality and $\tilde{J}_m$ is to be understood as its extension from $H_*^{-1,2}(\Omega)$ into $(H^{\bar{n}-1,2}(\Omega))^n \cap H_2(\Omega)$ (observe $P_{\frac{n+2}{n+1}}(u(.)\cdot\nabla u(.)) \in L^{\frac{n+2}{n+1}}((0,\tilde{T}),(L^{\frac{n+2}{n+1}}(\Omega))^n) \subset L^{\frac{n+2}{n+1}}((0,\tilde{T}),(H^{-1,2}(\Omega))^n) \subset L^{\frac{n+2}{n+1}}((0,T),H_*^{-1,2}(\Omega))$ as was proved in proposition IV.2.2 and in the proof of theorem IV.3.2). $(f(t),\tilde{J}_m\psi(t)) = (\tilde{J}_m P_{-1,2}f(t),\psi(t))$ with $\tilde{J}_m$ defined as before, $(\varphi,\tilde{J}_m\psi(0)) = (\tilde{J}_m\varphi,\psi(0))$. Since we have the estimate

$$\left|\int_\Omega \tilde{J}_m P_{\frac{n+2}{n+1}}(u(t)\cdot\nabla u(t))\,\bar{\tilde{\varphi}}\,dx\right| \le c(m)\|u(t)\cdot\nabla u(t)\|_{L^1(\Omega)}\cdot\|\tilde{\varphi}\|_{H_2(\Omega)},$$

$\tilde{\varphi} \in C_0^\infty(\Omega)$, $\nabla\cdot\tilde{\varphi} = 0$, we get

$$\|\tilde{J}_m P_{\frac{n+2}{n+1}}(u(t)\cdot\nabla u(t)\|_{H_2(\Omega)} \le c(m)\|\nabla u(t)\|_{(L^2(\Omega))^{n^2}}\|u(t)\|_{H_2}.$$

Thus $\tilde{J}_m P_{\frac{n+2}{n+1}}(u(t)\cdot\nabla(t)) \in L^2((0,T),H_2(\Omega))$, $\tilde{J}_m u' \in L^2((0,\tilde{T}),(L_2(\Omega))^n)$, $\tilde{J}_m u \in L^2((0,\tilde{T}),(H^{2,2}(\Omega))^n) \cap L^2((0,\tilde{T}),(\overset{\circ}{H}{}^{1,2}(\Omega))^m)$ and

$$\text{(IV.4.4)}\quad (\tilde{J}_m u)' + A_2(\tilde{J}_m u)' + \tilde{J}_m P_{\frac{n+2}{n+1}}(u\cdot\nabla u) = \tilde{J}_m P_{-1,2} f,$$

$$\tilde{J}_m u(0) = \tilde{J}_m \varphi,$$

as an equation in $L^2((0,\tilde{T}),H_2(\Omega))$.[1] In particular it follows that $A_2^{\frac{1}{2}}\tilde{J}_m u \in C^0([0,\tilde{T}],H_2(\Omega))$. The next part of the proof is like the proof of theorem III.2 in [SOW1]; the idea how to compare u^1,u^2 is due to Serrin [Ser2]. Taking (IV.4.4) for $u=u^1$ and $u=u^2$ we get because of

$$(\tilde{J}_m u^1, J_m u^2)' = ((\tilde{J}_m u^1)', J_m u^2) + (\tilde{J}_m u^1, (J_m u^2)') \text{ a.e. on } (0,\tilde{T})$$

(see auxiliary propositions (0.2.6) the equation

$$(\tilde{J}_m u^1(t), \tilde{J}_m u^2(t)) - (\tilde{J}_m \varphi^1, \tilde{J}_m \varphi^2) + 2\nu \int_0^{\tilde{T}} (\nabla \tilde{J}_m u^1, \nabla J_m u^2)\, dt +$$
$$+ \int_0^{\tilde{T}} (\tilde{J}_m P_{\frac{n+2}{n+1}}(u^1(t)\cdot\nabla u^1(t)), \tilde{J}_m u^2(t))\, dt +$$
$$+ \int_0^{\tilde{T}} (\tilde{J}_m u^1(t), \tilde{J}_m P_{\frac{n+2}{n+1}}(u^2(t)\cdot\nabla u^2(t)))\, dt =$$
$$= \int_0^{\tilde{T}} (\tilde{J}_m P_{-1,2} f^1(t), \tilde{J}_m u^2(t))\, dt + \int_0^{\tilde{T}} (\tilde{J}_m u^1(t), \tilde{J}_m P_{-1,2} f^2(t))\, dt.$$

We have (cf. the proof of (IV.3.6))

$$(\tilde{J}_m P_{\frac{n+2}{n+1}}(u^1(t)\cdot\nabla u^1(t)), \tilde{J}_m u^2(t))$$
$$= -(u^1(t)\cdot u^1(t), \nabla \tilde{J}_m^2 u^2(t)), (\tilde{J}_m u^1(t), \tilde{J}_m P_{\frac{n+2}{n+1}}(u^2(t)\cdot\nabla u^2(t)))$$
$$= -(u^1(t), \tilde{J}_m^2 P_{\frac{n+2}{n+1}}(u^2(t)\cdot\nabla u^2(t)),$$

[1] (IV.4.4) is a special case of [So2, Lemma 3.2].

$$|(u^1(t)\cdot u^1(t),-\nabla\tilde{J}_m^2u^2(t))|$$

$$\leq \|\nabla\tilde{J}_m^2u^2(t)\|^2_{(L^2(\Omega))^{n^2}}\cdot\|u(t)\|_{H_2(\Omega)}\ \|u^1(t)\|_{(L^r(\Omega))^n}$$

$$\cdot\|\nabla u^1(t)\|_{(L^2(\Omega))^{n^2}}\cdot\|u(t)\|_{H^2(\Omega)},$$

$$\leq c\|u^2(t)\|^2_{(L^2(\Omega))^n}\|u^1(t)\|_{(L^r(\Omega))^n}\|\nabla u^1(t)\|_{(L^2(\Omega))^n},$$

$$(\tilde{J}_mu^1(t),\tilde{J}_mP_{\frac{n+2}{n+1}}(u^2(t)\cdot\nabla u^2(t))) =$$

$$= (\tilde{J}_m^2u^1(t),P_{\frac{n+2}{n+1}}(u^2(t)\cdot\nabla u^2(t)),$$

(IV.4.5) $$|(\tilde{J}_mu^1(t),\tilde{J}_mP_{\frac{n+2}{n+1}}(u^2(t)\cdot\nabla u^2(t)))|$$

$$\leq \|u^1(t)\|_{(L^{r_1}(\Omega))^n}\|P_{\frac{n+2}{n+1}}(u^2(t)\cdot\nabla u^2(t)\|_{L^{r_2}(\Omega)}$$

with $\frac{1}{r_1}=\frac{1}{r}$, $\frac{1}{r_2}=1-\frac{1}{r}$. Since with $\frac{1}{r_3}=\frac{1}{2}-\frac{1}{r}$

(IV.4.6) $$\|P_{\frac{n+2}{n+1}}(u^2(t)\cdot\nabla u^2(t))\|_{(L^{r_2}(\Omega))^n}$$

$$\leq c\|u^2(t)\|_{(L^r(\Omega))^n}\|\nabla u^2(t)\|_{(L^2(\Omega))^{n^2}}$$

we get

$$|(\tilde{J}_mu^1(t),\tilde{J}_mP_{\frac{n+2}{n+1}}(u^2(t)\cdot\nabla u^2(t)))|$$

$$\leq c\|u^1(t)\|_{(L^r(\Omega))^n}\|u^2(t)\|_{(L^{r_3}(\Omega))^n}\|\nabla u^2(t)\|_{(L^2(\Omega))^{n^2}}.$$

Now $\|u^2(t)\|_{(L^{r_3}(\Omega))^n} \leq c\|\nabla u^2(t)\|^{\frac{n}{r}}_{(L^2(\Omega))^{2}} \|u^2(t)\|^{1-\frac{n}{r}}_{(L^2(\Omega))^n}$ by the Nirenberg-Gagliardo inequality in 0.2. The estimate then is

$$|(\tilde{J}_m u^1(t), \tilde{J}_m P_{\frac{n+2}{n+1}}(u^2(t)\cdot\nabla u^2(t)))|$$

$$\leq c\|u^1(t)\|_{(L^r(\Omega))^n} \|\nabla u^2(t)\|^{1+\frac{n}{r}}_{(L^2(\Omega))^{n^2}} \|u(t)\|^{1-\frac{n}{r}}_{H_2(\Omega)}.$$

We have to give some explanations for (IV.4.5), (IV.4.6). Since $(I+\frac{1}{m}A_p)u = (I+\frac{1}{m}A_q)u$ if $u \in D(A_{\max(p,q)})$ we see that $(I+\frac{1}{m}A_p)^{-[\frac{n}{2}]-1} u = (I+\frac{1}{m}A_q)^{-[\frac{n}{2}]-1} u$ if $u \in H_{\max(p,q)}(\Omega)$. Here we can set $p = r \geq q = 2$. Moreover

$$\|(I+\tfrac{1}{m}A_p)^{-1}u\|_{H_p(\Omega)} \leq c(p),\ m \in \mathbb{N},\ p > 1.$$

This proves (IV.4.5). As for (IV.4.6) we have used that $P_q(L^2(\Omega))^n \cap P_p(L^p(\Omega))^n = P_{\max(p,q)}(L^{\max(p,q)}(\Omega))^n$ as was stated in our auxiliary propositions 0.2. Since

$$(I+\tfrac{1}{m}A_2)^{-1}u \to u,\ u \in H_2(\Omega),$$

and therefore

$$(I+\tfrac{1}{m}A_2)^{-[\frac{n}{2}]-1} u \to u,\ u \in H_2(\Omega)$$

we can apply the dominated convergence theorem (Observe that all our constants do not depend on m unless we have written c(m)). The result is

$$\int_0^{\tilde T} (\tilde J_m P_{\frac{n+2}{n+1}}(u^1(t)\cdot\nabla u^1(t)), \tilde J_m u^2(t))\, dt$$

$$\to \int_0^{\tilde T} (P_{\frac{n+2}{n+1}}(u^1(t)\cdot\nabla u^1(t)), u^2(t))\, dt =$$

$$= \int_0^{\tilde T} (u^1(t)\cdot\nabla u^1(t), u^2(t))\, dt,\ m\to\infty,$$

$$\int_0^{\tilde T} (\tilde J_m u_1(t), \tilde J_m P_{\frac{n+2}{n+1}}(u^2(t)\cdot\nabla u^2(t)))\, dt$$

$$\to \int_0^{\tilde T} (u_1(t), P_{\frac{n+2}{n+1}}(u^2(t)\cdot\nabla u^2(t))\, dt =$$

$$= \int_0^{\tilde T} (u_1(t), P_{\frac{n+2}{n+1}}(u^2(t)\cdot\nabla u^2(t))\, dt,\ m\to\infty,$$

where the last two equalities are due to the property of the P_q, P_p just mentioned and the equality $\langle P_p u, P_q v\rangle = \langle u,v\rangle$ if $u \in P_p(L^p(\Omega))^n$, $v \in P_q(L^q(\Omega))^n$ and $(L^q(\Omega))^n \in (L^{p/(p-1)}(\Omega))^n$, for which we refer also to 0.2 $\langle .,.\rangle$ is the L^p-L^q duality. Thus we get

$$\text{(IV.4.7)}\quad (u^1(t),u^2(t)) - (\varphi_1,\varphi_2) + 2\nu \int_0^{\tilde T} (\nabla u^1(\sigma),\nabla u^2(\sigma))\, d\sigma + \int_0^{\tilde T} (u^1(t)\cdot\nabla u^1(t), u^2(t))\, dt + \int_0^{\tilde T} (u^1(t), u^2(t)\cdot\nabla u^2(t))\, dt = \int_0^{\tilde T} (f^1(t),u^2(t))\, dt + \int_0^{\tilde T} (u^1(t),f^2(t))\, dt.$$

Now we set $\tilde u = u^1 - u^2$. We have by the energy inequality for u^1, the energy inequality for u^2 and (IV.4.7)

$$\|u^1(t)-u^2(t)\|^2_{(L^2(\Omega))^n} + 2\nu \int_0^{\tilde{T}} \|\nabla u^1(t)-\nabla u^2(t)\|^2_{(L^2(\Omega))^{n^2}} dt$$

$$= \|u^1(t)\|^2 + \|u^2(t)\|^2 - 2(u^1(t),u^2(t)) + 2\nu \int_0^{\tilde{T}} \|\nabla u^1(t)\|^2_{(L^2(\Omega))^{n^2}} dt$$

$$+ 2\nu \int_0^{T} \|\nabla u^2(t)\|^2_{(L^2(\Omega))^{n^2}} dt - 4\nu \int_0^{\tilde{T}} (\nabla u^1(\sigma),\nabla u^2(\sigma))\, d\sigma$$

$$\leq \|\varphi^1\|^2 + \|\varphi^2\|^2 - 2(u^1(t),u^2(t)) - 4\nu \int_0^{\tilde{T}} (\nabla u^1(\sigma),\nabla u^2(\sigma))\, d\sigma$$

$$+ 2 \int_0^{\tilde{T}} (f^1(t),u^1(t))\, dt + 2 \int_0^{\tilde{T}} (f^2(t),u^2(t))\, dt,$$

$$= \|\varphi^1\|^2_{H_2(\Omega)} + \|\varphi^2\|^2_{H_2(\Omega)} - 2(\varphi_1,\varphi_2) +$$

$$+ 2 \int_0^{T} (u^1(t)\cdot\nabla u^1(t),u_2(t))\, dt$$

$$+ 2 \int_0^{\tilde{T}} (u^1(t),u^2(t)\cdot\nabla u^2(t))\, dt$$

$$+ 2 \int_0^{\tilde{T}} (f^1(t),u^1(t))\, dt - 2 \int_0^{\tilde{T}} (f^1(t),u^2(t))\, dt$$

$$- 2 \int_0^{\tilde{T}} (u^1(t),f^2(t))\, dt + 2 \int_0^{\tilde{T}} (f^2(t),u^2(t))\, dt,$$

$$= \|\varphi^1-\varphi^2\|^2_{H_2(\Omega)} + 2 \int_0^{\tilde{T}} (f_1(t)-f_2(t),u^1(t)-u^2(t))\, dt.$$

Since by partial integration (remember: u^1,u^2 are real)

$$\int_0^{\tilde{T}} ((u^1-u^2)\cdot\nabla(u^1-u^2),u^1)\, dt =$$

$$= \int_0^{\tilde{T}} (u^1\cdot\nabla u^1,u^1)\, dt + \int_0^{\tilde{T}} u^2\cdot\nabla u^2,u^1)\, dt$$

$$- \int_0^{\tilde{T}} (u^1\cdot\nabla u^2,u^1)\,dt - \int_0^{\tilde{T}} (u^2\cdot\nabla u^1,u^1)\,dt,$$

$$= \int_0^{T} (u^1\cdot\nabla u^1,u^1)\,dt + \int_0^{T} (u^2\cdot\nabla u^2,u^1)\,dt$$

we arrive at the inequality

$$\|u^1(t)-u^2(t)\|^2_{(L^2(\Omega))^n} + 2\nu \int_0^{\tilde{T}} \|\nabla(u^1(t)-u^2(t))\|^2\,dt$$

$$\le \|\varphi^1-\varphi^2\|^2_{H_2(\Omega)} + 2\int_0^{\tilde{T}} ((u^1-u^2)(t)\cdot\nabla(u^1-u^2)(t),u^1(t))\,dt +$$

$$+ 2\int_0^{\tilde{T}} (f^1(t)-f^2(t),u^1(t)-u^2(t))\,dt.$$

Using now the estimate (IV.3.6), (IV.3.7), (IV.3.8) with $\psi(t) = u^1(t)-u^2(t)$, $u(t) = u^1(t)$, $v(t) = u^1(t)-u^2(t)$ and the equality

$$((u^1-u^2)(t)\cdot\nabla(u^1-u^2)(t),u^1(t)) =$$
$$= -(u^1(t)\cdot(u^1(t)-u^2(t)),\nabla(u^1-u^2(t))$$

we get the inequalities of our theorem (Observe that $\frac{1}{s}=\frac{1}{2}-\frac{n}{2r}$, $\frac{s-2}{2s}=\frac{1}{2}-\frac{1}{s}=\frac{n}{2r}$, $s=\frac{2r}{n-r}$).

Now let $\varphi^1=\varphi^2$, $f^1=f^2$. As for the case $r>n$ the uniqueness is derived as follows: We have (observe that $1-\frac{1}{s}=\frac{1}{2}+\frac{n}{2r}$),

$$\left(\int_0^t \|u^1(\sigma)\|^s_{(L^r(\Omega))^n}\|u^1(\sigma)-u^2(\sigma)\|^2_{(L^2(\Omega))^n}\,dt\right)^{\frac{1}{s}}\cdot$$

$$\cdot\left(\int_0^t \|\nabla u^1(\sigma)-\nabla u^2(\sigma)\|^2_{(L^2(\Omega))^{n^2}}\,d\sigma\right)^{\frac{1}{2}+\frac{n}{2r}}$$

$$\le c(\varepsilon)\int_0^t \|u^1(\sigma)\|^s_{(L^r(\Omega))^n}\|u^1(\sigma)-u^2(\sigma)\|^2_{(L^2(\Omega))^n}\,d\sigma +$$

$$+ \varepsilon\int_0^t \|\nabla u^1(\sigma)-\nabla u^2(\sigma)\|^2_{(L^2(\Omega))^{n^2}}\,d\sigma,$$

$$\|u^1(t)-u^2(t)\|^2_{(L^2(\Omega))^{n^2}}$$

$$\leq c(\varepsilon) \int_0^t \|u^1(\sigma)\|^2_{(L^r(\Omega))^n} \|u^1(t)-u^2(t)\|_{(L^2(\Omega))^n} \, dt.$$

Since $\|u^1(.)\|^s_{(L^r(\Omega))^n} \in L^1((0,T))$ the assertion follows by Gronwall's inequality. In the third case $(s=2)$ we have

$$\int_0^t \|u^1(\sigma)\|_{(L^\infty(\Omega))^n} \|u^1(\sigma)-u^2(\sigma)\|_{(L^2(\Omega))^n} \|\nabla u^1(\sigma)-\nabla u^2(\sigma)\|_{(L^2(\Omega))^{n2}}$$

$$\leq c(\varepsilon) \int_0^t \|u^1(\sigma)\|^2_{(L^\infty(\Omega))^n} \|u^1(\sigma)-u^2(\sigma)\|^2_{(L^2(\Omega))^n} \, d\sigma +$$

$$+ \varepsilon \int_0^t \|\nabla u^1(\sigma)-\nabla u^2(\sigma)\|^2_{(L^2(\Omega))^{n^2}} \, d\sigma$$

and since $\|u^1(t)\|^2_{L^\infty(\Omega)} \in L^1((0,T))$ Gronwall's inequality completes the proof. In the case $r=n$ it cannot be shown from the inequality of our theorem that $u^1=u^2$ unless $\|u^1(t)\|_{(L^n(\Omega))^n} \leq \varepsilon_o$ for a sufficiently small ε_o, $0<t<T$. To overcome this difficulty we additionally introduce the assumption that $u^1 \in C^o([0,T], (L^n(\Omega))^n)$. Then there is a sequence u_k with

$$u_k^1 \in C^o([0,T], H_n(\Omega))$$

$$u_k^{(1)}(t) \in D(A_n)$$

$$A_n u_k^1(.) \in C^o([0,T], H_n(\Omega)),$$

$$u_k^1 \to u^1 \text{ in } C^o([0,T], H_n(\Omega)),\ k \to \infty.$$

Let $\varepsilon > 0$. Writing $u^1 = u^1 - u_k^1 + u_k^1$ we see that u^1 is decomposed into $u^1 = \bar{u}^1 + \bar{\bar{u}}^1$ with $\bar{u}^1 = u^1 - u_k^1$, $\bar{\bar{u}}^1 = u_k^1$,

$$\bar{\bar{u}}^1 \in D(A_n),$$

$$\|u^1-u_k\|_{H_n(\Omega)} = \|\bar{u}^1\|_{H_n(\Omega)} < \varepsilon \text{ if } k \geq k(\varepsilon).$$

By Sobolev $D(A_n) \subset (C^o(\bar{\Omega}))^n$ with a continuous imbedding. Thus we get as in the proof of (IV.3.7) the inequalities

$$|((u^1(t)-u^2(t))\cdot\nabla(u^1-u^2)(t),u^1(t))|$$

$$\leq \|\bar{u}^1(t)\|_{(L^n(\Omega))^n}\|\nabla(u^1-u^2)(t)\|^2_{(L^2(\Omega))^{n^2}} +$$

$$+ \|\bar{u}^2(t)\|_{(C^o(\bar{\Omega}))^n}\|u^1(t)-u^2(t)\|_{H_2(\Omega)}\|\nabla(u^1-u^2)(t)\|^2_{(L^2(\Omega))^{n^2}},$$

$$\leq \varepsilon\|\nabla(u^1-u^2)(t)\|^2_{(L^2(\Omega))^{n^2}} +$$

$$+ \varepsilon'\|\bar{u}^2(t)\|^2_{(C^o(\bar{\Omega}))^n}\|\nabla(u^1-u^2)(t)\|^2_{(L^2(\Omega))^{n^2}} +$$

$$+ c(\varepsilon')\|u^1(t)-u^2(t)\|^2_{H_2(\Omega)}$$

for any $\varepsilon > 0$ if $k \geq k(\varepsilon)$ and any $\varepsilon' > 0$. Thus we arrive at

$$\|u^1(t)-u^2(t)\|^2_{H_2(\Omega)} + 2\nu \int_0^t \|\nabla u^1(\sigma)-\nabla u^2(\sigma)\|^2_{L^2(\Omega)} \, d\sigma$$

$$\leq \|\varphi^1-\varphi^2\|^2_{H_2(\Omega)} + 2 \int_0^t (f^1(\sigma)-f^2(\sigma),u^1(\sigma)-u^2(\sigma)) \, d\sigma$$

$$+ \varepsilon \int_0^t \|\nabla(u^1-u^2)(\sigma)\|^2_{(L^2(\Omega))^{n^2}} \, d\sigma$$

$$+ \varepsilon' \sup_{0\leq\sigma\leq t} \|\bar{u}^2(\sigma)\|^2_{(C^o(\bar{\Omega}))^n} \int_0^t \|\nabla(u^1-u^2)(\sigma)\|^2_{(L^2(\Omega))^{n^2}} \, d\sigma +$$

$$+ c(\varepsilon') \int_0^t \|u^1(\sigma)-u^2(\sigma)\|^2_{H^2(\Omega)} \, d\sigma.$$

Taking first k sufficiently large (this means that ε is small) and then ε' sufficiently small we see that Gronwall's inequality gives $u^1 = u^2$ if $\varphi^1 = \varphi^2$, $f^1 = f^2$. Our theorem is proved. □

It may be noted that the assumption that u^1 fulfills the energy inequality, is redundant since from theorem IV.3.1 we know already that u^1 fulfills the energy equality (IV.3.2) everywhere, but we have made this assumption nevertheless to point out that only the energy inequality (IV.4.1) is needed. Moreover theorem IV.4.1 immediately gives the result that

(IV.4.8) $L^s((0,T),(L^r(\Omega))^n)$, $\frac{2}{s}+\frac{n}{r}=1$, $2 \leq s < +\infty$, $n < r \leq +\infty$,

and

(IV.4.9) $C^0([0,T],(L^n(\Omega))^n)$

are uniqueness classes for weak solutions of the Navier-Stokes equations in the sense of definition IV.1.1. The first class has been shown to be a uniqueness class by Lions [Li, 1.6., p. 84] too. The second result goes back to [SOW1, W8]. It may be left to the reader that the stronger condition

(IV.4.10) $L^s((0,T),(L^r(\Omega))^n)$, $\frac{2}{s}+\frac{n}{r}<1$, $2 \leq s < +\infty$, $n < r \leq +\infty$

also yields a uniqueness class, and in particular that weak solutions belonging to this class fulfill the energy equality everywhere (in the sense as has been pointed out in theorem IV.4.1). Thus it remains to deal with the class

$$L^\infty((0,T),(L^n(\Omega))^n)$$

further on. We will show that uniqueness also holds within this class (we even prove a stronger result); the proof however has a conceptual disadvantage for it heavily relies on the reconstruction of the weak solution as a strong one locally in time. By strong solution here we mean a solution in the sense of theorem III.3.6. Following these lines we later on give a different (local in time)-uniqueness proof for weak solutions in $L^4((0,T),(L^4(\Omega))^n)$ which is however limited to the dimensions 3,4.

Theorem IV.4.2: Let $\varphi \in H_n(\Omega)$, let

$$f \in C^{\frac{1}{p}}([0,T],(L^p(\Omega))^n) \cap C^0((0,T],(H^{\frac{1}{p},p}(\Omega))^n)$$

for some $p > n$. Let u^1, u^2 be weak solutions of the Navier-Stokes equations over $(0,T) \times \Omega$ in the sense of definition IV.1.1 with the same boundary values $u^i(0) = \varphi$, $u^i|\partial\Omega = 0$, $i = 1,2$, and the same inhomogeneous term f. Let u^1 fulfill the energy inequality everywhere, i.e.

$$\text{(IV.4.11)}\quad \|u^1(t)\|_{(L^2(\Omega))^n} + 2\nu \int_s^t \|\nabla u^1(\sigma)\|^2_{(L^2(\Omega))^{n^2}} d\sigma$$

$$\leq \|u^1(s)\|^2_{(L^2(\Omega))^n} + 2 \int_s^t (f(\sigma),u(\sigma))\, d\sigma$$

for all s,t, $0 \leq s \leq t \leq T$. Let

$$\text{(IV.4.12)}\quad u^2 \in L^\infty((0,T),(L^n(\Omega))^n).$$

Then $u^1(t) = u^2(t)$ on $[0,T]$.

Proof: By definition IV.1.1 $u:[0,T] \to H_2(\Omega)$ is weakly continuous. Since $u(t) \in (L^n(\Omega))^n$ and therefore $u(t) \in H_n(\Omega)$ a.e. on $(0,T)$, and since (IV.4.12) holds, we get that

$$u:[0,T] \to H_n(\Omega)$$

is weakly continuous. In particular this means that $u(t) \in H_n(\Omega)$ on $[0,T]$. Theorem (IV.3.1) shows that u^2 even fulfills the energy equality everywhere. In particular u^2 fulfills

$$(IV.4.13)\quad \|u^2(t)\|_{(L^2(\Omega))^n} + 2\nu \int_s^t \|\nabla u^2(\sigma)\|_{(L^2(\Omega))^{n^2}} d\sigma \le \|u^1(s)\| + 2\int_s^t (f(u(\sigma)),u(\sigma))\, d\sigma$$

for all s,t with $0 \le s \le t \le T$, and this we only need.
We construct a third solution of the Navier-Stokes equations with data φ, f, according to theorem III.3.6 on some interval $[0,T^*]$, $0 < T^* \le T$. u^3 is in $C^0([0,T^*],(L^n(\Omega))^n)$ and we have

$$u^{3'} - \nu\Delta u^3 + P(u^3 \cdot \nabla u^3) = Pf$$

on every (ε, T^*), $0 < \varepsilon < T^*$, as an equation in $C^0([\varepsilon,T],$ $(L^p(\Omega))^n)$. This equation can be multiplied by u^3, all partial integrations can be carried through in order to see that

$$(IV.4.14)\quad \|u^3(t)\|^2_{(L^2(\Omega))^n} + 2\nu \int_s^t \|\nabla u^3(\sigma)\|^2_{(L^2(\Omega))^{n^2}} d\sigma = \|u^3(s)\|_{(L^2(\Omega))^n} + 2\int_s^t (f(\sigma),u^3(\sigma))\, d\sigma$$

for all s,t, $\varepsilon \le s \le t \le T^*$. We can let ε tend to 0 since $u^3 \in C^0([0,T^*],(L^n(\Omega))^n)$, and we get that (IV.4.14) is fulfilled for all s,t, $0 \le s \le t \le T$. Thus u^3 is a weak solution of the Navier-Stokes equations over $(0,T^*)\times\Omega$ as in definition IV.1.1 which fulfills the energy inequality everywhere. Theorem IV.4.1 now shows that

$$u^2(t) = u^3(t) \text{ on } [0,T^*]$$

(cf. theorem IV.1.1). Since (IV.4.13) holds a second application of theorem IV.4.1 yields

$$u^3(t) = u^1(t) \text{ on } [0,T^*]$$

and thus

$$u^1(t) = u^2(t) \text{ on } [0,T^*].$$

Now we set

$$t_o = \sup\{t \mid 0 \le t \le T,\ u^1(t) = u^2(t)\}.$$

Assuming that $t_o < T$ we get a contradiction: From the weak continuity of u^1, u^2 we obtain $u^1(t_o) = u^2(t_o)$. We already know that $u^2(t_o) \in H_n(\Omega)$, thus $u^1(t_o) \in H_n(\Omega)$ and we can apply the same argument as before to see that $u^1(t) = u^2(t)$ on some interval $[t_o, t_o + T^{**}]$, $t_o < t_o + T^{**} \le T$. This completes the proof. □

We want to make a few remarks about the assumptions of the preceding theorem. First of all the theorem remains true if $f \in L^n((0,T),(H^{-1,n}(\Omega))^n)$, for a proof see [SOW1]. If we assume that the energy inequality (IV.4.11) holds for u^2 too then the assumption (IV.4.12) may be weakened, namely we have to assume only that $u^2(t) \in (L^n(\Omega))^n$ for all t, $0 \le t < T$. This is easily seen by the proof of theorem IV.4.2: We can construct then $u^3(t)$ on any interval $[t_o, t_o+\varepsilon]$ with $0 \le t_o < t_o+\varepsilon \le T$ with $u^2(t_o)$ as initial value. Theorem IV.4.2 moreover shows that at least under the assumptions on φ, f having been made in the preceding theorem the function space

$$L^\infty((0,T),(L^n(\Omega))^n)$$

is a uniqueness class for weak solutions.

From our uniqueness theorems and the construction of local strong solutions it will become clear that a weak solution constructed via the Galerkin approximation procedure at least locally in t can be reconstructed as a strong one provided the weak one is almost everywhere in $(H^{2,p}(\Omega))^n$ for some $p > 1, \frac{n}{3}$ or in $(L^{n+\delta}(\Omega))^n$ for some $\delta > 0$ or in $(L^n(\Omega))^n$. We start with the first case, where we want to give a proof being independent from theorem IV.4.1.

Theorem IV.4.3: Let $\varphi \in (H^{2,p}(\Omega))^n \cap (\overset{o}{H}{}^{1,p}(\Omega))^n \cap H_p(\Omega) = D(A_p)$ for some $p > 1, \frac{n}{3}$. Let $f \in C^{o,1}([0,T],(L^p(\Omega))^n) \cap L^{\frac{n+2}{n+1}}((0,T),(L^{\frac{n+2}{n+1}}(\Omega))^n)$ for some $T > 0$, let u^1 be a weak solution of the Navier-Stokes equations over $(0,T)\times\Omega$ in the sense of definition IV.1.1 with boundary values $u(0) = \varphi$, $u|\partial\Omega = 0$ and inhomogeneous term f. Let u^2 be a solution of

$$u' + A_p u + M(u) = Pf,$$
$$u(0) = \varphi$$

in the sense of theorem III.3.1 over some interval $[0,T^*]$ with $0 < T^* \leq T$, which in particular means that

$$u^2 \in C^1([0,T^*],H_p(\Omega))$$
$$u^2(t) = D(A_p),\ 0 \leq t \leq T^*,$$
$$A_p u^2(.) \in C^o([0,T^*],H_p(\Omega)).$$

Finally let

$$u^1 \cdot u^1 \in L^2((0,T),(L^2(\Omega))^{n^2}).$$

Then $u^1(t) = u^2(t)$ on $[0,T^*]$.

Proof: The proof is done by the use of cut off functions. Let us set for any real $w \in (H^{1,q}(\Omega))^n$ with some $q > 1$ and any $K \in \mathbb{N}$

$$w_{l,K}(x) = \begin{cases} K, & w_l(x) \geq K, \\ w_l(x), & -K \leq w_l(x) \leq K, \\ -K, & w_l(x) \leq -K, \end{cases}$$

where w_l is the l-th component of w. Then the vector w_K with components $w_{1,K},\ldots,w_{n,K}$ is also in $(H^{1,q}(\Omega))^n$. If $w \in (\overset{o}{H}{}^{1,q}(\Omega))^n$ then $w_K \in (\overset{o}{H}{}^{1,q}(\Omega))^n$. Moreover we have

(IV.4.15) $\|w_K\|_{(L^q(\Omega))^n} \leq \|w\|_{(L^q(\Omega))^n}$,

(IV.4.16) $\begin{cases} (\frac{\partial}{\partial x_i} w_{l,K})(x) = 0 \text{ if } |w_l(x)| \quad K, \; (\frac{\partial}{\partial x_i} w_{l,K})(x) = (\frac{\partial}{\partial x_i} w_l)(x) \\ \qquad\qquad\qquad\qquad\qquad\qquad \text{otherwise a.e. on } \Omega, \\ \|\nabla w_K\|_{(L^q(\Omega))^{n^2}} \leq \|\nabla w\|_{(L^q(\Omega))^{n^2}}, \end{cases}$

(IV.4.17) $\|w - w_k\|_{(H^{1,q}(\Omega))^n} \to 0$ for $K \to \infty$

(cf. 0.2 for these well known facts). Unfortunately, if $\nabla \cdot w = 0$, then this not so with $\nabla \cdot w_K$ but

$$\|\nabla \cdot w_K\|_{L^q(\Omega)} \to 0 \text{ for } K \to \infty.$$

By theorem IV.3.2 we get

$$u^{1\prime} - \nu\Delta u^1 + u^1 \cdot \nabla u^1 + \nabla\pi^1 = f$$

as an equation in $L^{\frac{n+2}{n+1}}((0,T),(L^{\frac{n+2}{n+1}}(\Omega))^n)$; for this observe that by Sobolev

$$(W^{2-\frac{2(n+1)}{n+2},\frac{n+2}{n+1}}(\Omega))^n \subset (H^{1,\frac{n+2}{n+1}}(\Omega))^n \subset (H^{2,p}(\Omega))^n$$

for any $p > 1$. Of course we have

$$u^{2\prime} - \nu\Delta u^2 + u^2 \cdot \nabla u^2 + \nabla\pi^2 = f$$

in $C^0([0,T^*],(L^p(\Omega))^n)$ (cf. III.1). This gives

(IV.4.18) $(u^1 - u^2)' - \nu\Delta(u^1 - u^2) + u^1 \cdot \nabla(u^1 - u^2) + (u^1 - u^2) \cdot \nabla u^2 + \nabla\pi^1 - \nabla\pi^2 = 0$

as an equation in $L^{\frac{n+2}{n+1}}((0,T^*),(L^{\tilde{p}}(\Omega))^n)$ with $\tilde{p}=\min(\frac{n+2}{n+1},p)>1$. Multiplying both sides of (IV.4.18) by $(u^1-u^2)_K$ we get ($<.,.>$ is the duality between $(L^{\tilde{p}}(\Omega))^n$ and $(L^{\tilde{p}/(\tilde{p}-1)}(\Omega))^n$)

$$(IV.4.19)\quad <(u^1-u^2)',(u^1-u^2)_K> - \nu<\Delta(u^1-u^2),(u^1-u^2)_K>$$
$$+ <u^1\cdot\nabla(u^1-u^2),(u^1-u^2)_K> + <(u^1-u^2)\cdot\nabla u^2,(u^1-u^2)_K>$$
$$+ <\nabla\pi^1-\nabla\pi^2,(u^1-u^2)_K> = 0.$$

Let us consider the various terms in (IV.4.19) pointwise in t. By partial integration we get with (IV.4.16)

$$-\nu<\Delta(u^1-u^2)(t),(u^1-u^2)_K(t)> = \nu\|\nabla(u^1-u^2)_K(t)\|^2_{(L^2(\Omega))^n},$$

$$<u^1\cdot\nabla(u^1-u^2)(t),(u^1-u^2)_K(t)>$$
$$= \sum_{l=1}^{n}\int_\Omega\sum_{i=1}^{n} u_i^1(t)\frac{\partial(u^1-u^2)_l(t)}{\partial x_i}\cdot(u^1-u^2)_{l,K}(t)\,dx,$$
$$= -\sum_{i,l=1}^{n}\int_\Omega u_i^1(t)(u^1-u^2)_l(t)\cdot\frac{\partial}{\partial x_i}(u^1-u^2)_{l,K}(t)\,dx,$$
$$= -\frac{1}{2}\sum_{i,l=1}^{n}\int_\Omega u_i^1(t)\frac{\partial}{\partial x_i}(u^1-u^2)^2_{l,K}(t)\,dx = 0$$

$$\int_0^t <(u^1-u^2)'(\sigma),(u^1-u^2)_K(\sigma)>\,d\sigma = \|\,|(u-v)(t)|\,|(u-v)_K(t)|\,\|_{L^1(\Omega)}$$
$$-\frac{1}{2}\|(u-v)_K(t)\|^2_{(L^2(\Omega))^n}.$$

The last equation is shown by approximating $(u^1-u^2)'$ in $L^{\frac{n+2}{n+1}}((0,T),(L^{\tilde{p}}(\Omega))^n)$ and u^1-u^2 in $L^{\frac{n+2}{n+1}}((0,T),(L^{\tilde{p}}(\Omega))^n)$ by functions ψ_m, $m\in\mathbb{N}$, with

$$\psi_m \in C^1([0,T],(L^2(\Omega))^n), \ \psi_m(0)=0$$

(cf. 0.2) and taking into consideration that the $\psi_{m,k}$ are in $C^{0,1}([0,T],(L^2(\Omega))^n)$, $\psi_{m,K}(0)=0$, and that $\psi'_{m,K} \to (u^1-u^2)'_K$, $\psi_{m,K} \to (u^1-u^2)_K$, $m\to\infty$, in the norms as indicated above (cf. 0.2). This yields

$$\text{(IV.4.19)} \quad 2\| \, |(u-v)(t)| \, |(u-v)_K(t)| \, \|_{L^1(\Omega)} - \|(u-v)_K(t)\|^2_{(L^2(\Omega))^n}$$
$$+ 2\nu \int_0^t \|\nabla(u^1-u^2)_K(\sigma)\|^2_{(L^2(\Omega))^{n^2}} d\sigma$$
$$\le \int_0^t \|u^1(\sigma)-u^2(\sigma)\|^2_{(L^2(\Omega))^n} \|\nabla u^2(\sigma)\|_{C^0(\overline{\Omega})} d\sigma +$$
$$+ \left| \int_0^t \langle \nabla\pi^1(\sigma)-\nabla\pi^2(\sigma),(u^1-u^2)_K(\sigma)\rangle \, d\sigma \right| .$$

We have to deal with the term $\|\nabla u^2(\sigma)\|_{C^0(\overline{\Omega})}$; first we must see that it is finite and then we estimate it. f can be continued to $[0,\infty)$ as to be in $C^{0,1}([0,\tilde{T}],(L^p(\Omega))^n)$ for all $\tilde{T}$. Thus u^2 is the solution of the Navier-Stokes equations having already been constructed in theorem III.3.1. Thus $T^* < T(\varphi)$. For this solution we know from (III.3.11) that

$$\|\nabla u^2(t)\|_{C^0(\overline{\Omega})} \le \frac{c}{t^{1-\rho}}.$$

Thus (IV.4.19) is justified. We obtain

$$\int_0^t \langle \nabla\pi^1-\nabla\pi^2,(u^1-u^2)_K\rangle \, d\sigma$$
$$= \int_0^t \langle \nabla\pi^1-\nabla\pi^2+(u^1-u^2)',(u^1-u^2)_K\rangle \, d\sigma -$$
$$- \int_0^t \langle (u^1-u^2)',(u^1-u^2)_K\rangle \, d\sigma.$$

We have by theorem IV.3.3

$$|\langle\nabla\pi^1-\nabla\pi^2+(u^1-u^2)',(u^1-u^2)_K\rangle|$$

$$\leq \|-\nu\Delta u^1-\nu\Delta u^2+u^1\cdot\nabla u^1+u^2\cdot\nabla u^2\|_{(H^{-1,2}(\Omega))^n}\|\nabla(u^1-u^2)\|_{(L^2(\Omega))^{n^2}}.$$

Thus by the theorem of dominated convergence

$$\int_0^t \langle\nabla\pi^1-\nabla\pi^2+(u^1-u^2)',(u^1-u^2)_K\rangle\, d\sigma$$

$$\to \int_0^t ((u^1-u^2)',u^1-u^2)\, d\sigma,\ K\to\infty.$$

Since $(u^1-u^2)'\in L^2((0,T^*),H_*^{-1,2}(\Omega))$, $u^1-u^2\in L^2((0,T^*),H_*^{1,2}(\Omega))$ we get by our auxiliary propositions (0.2.7) that $\int_0^t ((u^1-u^2)',u^1-u^2)\, dt = \frac{1}{2}\|u^1(t)-u^2(t)\|_{H_2(\Omega)}$. This gives

$$\int_0^t \langle\nabla\pi^1-\nabla\pi^2,(u^1-u^2)_K\rangle\, d\sigma \to 0,$$

for $K\to\infty$. Since $\int_0^t \|\nabla(u^1-u^2)_K(\sigma)\|^2_{(L^2(\Omega))^{n^2}}\, d\sigma \to$ $\int_0^t \|\nabla(u^1-u^2)(\sigma)\|^2_{(L^2(\Omega))^{n^2}}\, d\sigma$,

$$2\|\,|(u^1-u^2)(t)|\,|(u^1-u^2)_K(t)|\,\|_{L^1(\Omega)} - \|(u^1-u^2)_K(t)\|^2_{(L^2(\Omega))^n}$$

$$\to \|(u^1-u^2)(t)\|^2_{(L^2(\Omega))^n}$$

for $K\to\infty$, we arrive at the inequality

$$\|(u^1-u^2)(t)\|^2_{(L^2(\Omega))^n} + 2\nu\int_0^t \|\nabla(u^1-u^2)(\sigma)\|^2_{(L^2(\Omega))^{n^2}}\, d\sigma$$

$$\leq \int_0^t \|u^1(\sigma)-u^2(\sigma)\|^2_{(L^2(\Omega))^n}\frac{c}{\sigma^{1-\rho}}\, d\sigma.$$

Now Gronwall's inequality yields the result requested. □

<u>Remark:</u> This result may be also proved with the aid of theorems IV.3.3, IV.4.1 since from the first theorem it follows that u^1 even fulfills the energy inequality everywhere and since by Sobolev it follows from $u^2 \in C^o([0,T^*],(H^{2,p}(\Omega))^n)$ that $u^2 \in C^o([0,T^*],(L^{n+\varepsilon}(\Omega))^n)$ for some $\varepsilon > 0$; moreover u^2 is a weak solution over $(0,T^*)\times\Omega$ in the sense of definition IV.1.1 since by Sobolev $C^o([0,T^*],(\overset{o}{H}{}^{1,2}(\Omega))^n\ H_2(\Omega))$ is continuously imbedded into $C^o([0,T^*],(H^{2,p}(\Omega))^n)$ if $n \geq 4$; for all n this can be concluded in the way we have proved theorem IV.4.3, namely multiply $u^{2\prime} - \nu\Delta u^2 + u^2\cdot\nabla u^2 + \nabla\pi^2 = f$ scalarly by u^2_K; thus again by theorem IV.3.3 all assumptions of theorem IV.4.1 are fulfilled on $[0,T^*]$ (instead on $[0,T]$). But we have preferred it to present a different method.

Now we continue to compare weak solutions with local strong solutions.

<u>Theorem IV.4.4: Let</u> $\varphi \in H_p(\Omega)$ <u>for some</u> $p > n$. <u>Let</u> $f \in C^{1/p}([0,T], H_p(\Omega)) \cap C^o([0,T])$ <u>for some</u> $T > 0$. <u>Let</u> u^1 <u>be a weak solution of the Navier-Stokes equations over</u> $(0,T)\times\Omega$ <u>in the sense of definition</u> IV.1.1 <u>with boundary values</u> $u(0) = \varphi$, $u|\partial\Omega = 0$ <u>and inhomogeneous term</u> f. <u>Let</u> u^2 <u>be a solution of</u>

$$u' + A_p u + M(u) = Pf,$$
$$u(0) = \varphi$$

<u>in the sense of theorem</u> III.3.4 <u>over some interval</u> $[0,T^*]$ <u>with</u> $0 < T^* \leq T$, <u>which in particular means that</u>

$$u^2 \in C^1((0,T^*],H_p(\Omega)),$$

$$u^2(t) \in D(A_p),\quad 0 < t \leq T^*,$$

$$A_p u^2(.) \in C^o((0,T^*],H_p(\Omega)),$$

$$u^2 \in C^o([0,T^*],H_p(\Omega)),$$

$$.^{1-\rho} A_p^{1-\rho} u^2(.) \in L^\infty((0,T^*),H_p(\Omega))$$

for some $\rho \in (0,1)$ which is choosen in such a way that $\|\nabla v\|_{(C^0(\overline{\Omega}))^{n^2}} \le c\|A_p^{1-\rho}v\|$ for all $v \in D(A_p^{1-\rho})$. Finally we assume that u^1 fulfills the energy inequality.

$$\text{(IV.4.20)}\quad \|u^1(t)\|^2_{(L^2(\Omega))^n} + 2\nu \int_0^t \|\nabla u^1(\sigma)\|^2_{(L^2(\Omega))^{n^2}}\, d\sigma \le \|\varphi\|^2_{H_2(\Omega)} + 2\int_0^t (f(\sigma),u(\sigma))\, d\sigma$$

for all t, $0 \le t \le T$. Then

$$u^1(t) = u^2(t)$$

on $[0,T^*]$.

Proof: The proof is as outlined in the preceding remark. The proof that u^2 is a weak solution can be given as was done for u^3 in the proof of theorem IV.4.2. □

Now we state the theorem corresponding to theorem III.3.5 and to theorem III.3.6.

Theorem IV.4.5: I. Let $n=3$, let $\varphi \in D(A_2^{\frac{1}{4}})$. Let $f \in (C^{\frac{1}{4}}([0,T],H_2(\Omega)) \cap C^0([0,T],(H^{\frac{1}{2},2}(\Omega))^3)$ for some $T>0$. Let u^1 be a weak solution of the Navier-Stokes equations over $(0,T)\times\Omega$ as in definition IV.1.1 with boundary values $u(0)=\varphi$, $u|\partial\Omega = 0$ and inhomogeneous term f. Let u^2 be a solution of

$$\begin{aligned} u' + A_2u + M(u) &= P_2f \\ u(0) &= \varphi \end{aligned}$$

in the sense of theorem III.3.5 over some interval $[0,T^*]$ with $0 < T^* \le T$ which in particular means that

$$u^2 \in C^1((0,T^*],H_2(\Omega)),$$

$$u^2(t) \in D(A_2),\ 0 < t \leq T^*,$$

$$A_2 u^2(.) \in C^0((0,T^*],H_2(\Omega)),$$

$$u^2(t) \in D(A_2^{1/4}),\ 0 \leq t \leq T^*,\ A_2^{1/4} u^2(.) \in C^0([0,T^*],H_2(\Omega)),$$

$$\cdot^{\frac{1}{2}} A_2^{\frac{3}{4}} u^2(.) \in L^\infty((0,T^*),H_2(\Omega)).$$

Finally we assume that u^1 fulfills the energy inequality

$$\|u^1(t)\|^2_{(L^2(\Omega))^3} + 2\nu \int_0^t \|\nabla u^1(\sigma)\|^2_{(L^2(\Omega))^3} d\sigma$$

$$\leq \|\varphi\|^2_{H_2(\Omega)} + 2 \int_0^t (f(\sigma),u(\sigma))\, d\sigma$$

for all t, $0 \leq t \leq T$. Then

$$u^1(t) = u^2(t)$$

on $[0,T^*]$.

II. Let $\varphi \in H_n(\Omega)$. Let f be as in theorem IV.4.5 for some $T > 0$. Let u^1 be a weak solution of the Navier-Stokes equations over $(0,T) \times \Omega$ as in definition IV.1.1 with boundary values $u(0) = \varphi$, $u|\partial\Omega = 0$ and inhomogeneous term f. Let u^2 be a solution of

$$u' + A_n u + M(u) = P_n f$$
$$u(0) = \varphi$$

in the sense of theorem III.3.6 over some interval $[0,T^*]$ with $0 < T^* \leq T$, which in particular means that

$$u^2 \in C^1((0,T^*],H_p(\Omega)),$$

$$u^2(t) \in D(A_p),\ 0 < t \leq T^*,$$

$A_p u^2(.) \in C^o((O,T^*],H_p(\Omega))$,

$u^2 \in C^o([O,T^*],H_n(\Omega))$,

$A_n^{\frac{1}{4}} \frac{1}{4} u^2(.) \in L^\infty((O,T^*),H_n(\Omega))$.

Finally we assume that u^1 _fulfills the energy inequality_ (IV.4.20). _Then_

$u^1(t) = u^2(t)$

_on_ $[O,T^*]$.

Proof: The proof of part II of the present theorem has been given in the proof of theorem IV.4.2. As for part I. we refer to the remark preceding theorem IV.4.4. The proof that u^2 is a weak solution can be given as was done for u^3 in the proof of theorem IV.4.2 since by theorem III.2 and by Sobolev $C^o([O,T^*], D(A_2^{1/4}))$ is continuously imbedded in $C^o([O,T^*],(L^n(\Omega))^n)$ for $n=3$. □

§ 5. Regularity of Weak Solutions. Leray's Structure Theorem

We want to show in this paragraph that the uniqueness class

$$L^s((O,T),L^r(\Omega)), \ \frac{n}{r}+\frac{2}{s}=1, \ n<r<+\infty, \ 2<s<+\infty,$$

having been studied after the proof of theorem IV.4.1 is also a regularity class for low dimensions, i.e. any weak solution u from this class is regular in the sense that it is in $C^o((O,T],(H^{2,p}(\Omega))^n)$ for $p>n$ and under suitable assumptions on φ and f.

This has been proved partly by Sohr [So1]; the idea however to deal with the "critical quantity" $\frac{n}{r}+\frac{2}{s}$ goes back to Serrin [Ser1] who showed that under the stronger condition $\frac{n}{r}+\frac{2}{s}<1$, $n<r<+\infty$, $2<s<+\infty$, any weak solution is of class C^∞ in Ω if f is so; this result can be somewhat refined ([W7],[W2]). A complete result for $\Omega=\mathbb{R}^n$, $\frac{n}{r}+\frac{2}{s}=1$, $n<r<+\infty$, $2<s<+\infty$, and for any dimension was given by Fabes, John, and Riviere [FJR]. Moreover we will prove that $C^o([0,T],(L^n(\Omega))^n)$ is a regularity class; also certain subspaces of $L^\infty((0,T),L^n(\Omega))^n)$ are regularity classes; the first result is due to [W8], [W9], the second one to [So]. Finally we prove that the stability of weak solutions in the uniqueness class $L^\infty((0,T),(L^n(\Omega))^n)$ implies that $L^\infty((0,T),(L^n(\Omega))^n)$ is a regularity class; this result is due to [W4]. The results concerning $L^\infty((0,T),(L^n(\Omega))^n)$ or $C^o([0,T],(L^n(\Omega))^n)$ are valid for arbitrary n.

As for the class $L^\infty((0,T),(L^n(\Omega))^n)$ we were not able to show that this function space is a regularity class; however, we can show that a weak solution being in $L^\infty((0,T),(L^n(\Omega))^n)$ has at most countably many points in $[0,T]$ where $u(t)$ is not a member of $(H^{2,p}(\Omega))^n$ $(p>n)$. We will also show that any weak solution with energy inequality in dimensions $n=3,4$ has the property that the set of points $t\in[0,T]$ where $u(t)\notin(H^{2,p}(\Omega))^n$ has measure 0.

<u>Theorem IV.5.1:</u> <u>Let</u> $\varphi\in H_p(\Omega)$ <u>for some</u> $p>n$, <u>let</u> $f\in C^{\frac{1}{2p}}([0,T],(L^p(\Omega))^n)\cap C^o((0,T],(H^{\frac{1}{p},p}(\Omega))^n)$ <u>for all</u> $T>0$. <u>Let</u> $\tilde{u}$ <u>be a weak solution of the Navier-Stokes equations over</u> $(0,T)\times\Omega$ <u>for every</u> $T>0$ <u>with boundary values</u> $\tilde{u}(0)=\varphi$, $\tilde{u}|\partial\Omega=0$ <u>in the sense of definition</u> IV.1.1. <u>Let</u>

$$\tilde{u}\in L^s((0,T),(L^r(\Omega))^n),\ T>0,$$

<u>for some</u> r,s <u>with</u> $\frac{n}{r}+\frac{2}{s}=1$, $n<r\le+\infty$, $2\le s<+\infty$. <u>Then</u>

$$\tilde{u}\in C^o((0,T],(H^{1,2}(\Omega))^n)$$

for all $T > 0$ _and consequently_

$$\tilde{u} \in C^0([0,T], (L^n(\Omega))^n)$$

for all $T > 0$ _if_ $n \leq 4$; _thus for_ $n \leq 4$ _we get_

$$\tilde{u} \in C^0((0,T], (H^{2,p}(\Omega))^n).$$

Proof: Let us take the strong solution u of the Navier-Stokes equations from theorem III.3.4. We know already that

$$\tilde{u}(t) = u(t) \text{ on } [0,T(\varphi))$$

(cf. theorem IV.4.4). We have to show that $T(\varphi) = +\infty$. In particular we have

$$u(t) \in D(A_p),\ t \in (0,T(\varphi))$$

$$A_p u \in C^0((0,T(\varphi)), H_p(\Omega)),$$

$$\text{(IV.5.1)}\quad u' + A_p u + M(u) = P_p f.$$

As we have already mentioned in the end of III.4, theorem III.3.4 remains valid in any number of space dimensions. It is also clear from these remarks that due to [Gi1] the operator A_2 is selfadjoint in $H_2(\Omega)$. Thus multiplying (IV.5.1) scalarly by u' and $A_2 u$ we get

$$\|A_2^{\frac{1}{2}} u(t)\|^2_{H_2(\Omega)} + \int_\varepsilon^t \|A_2 u(\sigma)\|^2_{H_2(\Omega)}\, d\sigma$$

$$\leq c\left(\int_\varepsilon^t \|M(u(\sigma))\|^2_{H_2(\Omega)}\, d\sigma + \int_\varepsilon^t \|P_2 f(\sigma)\|^2_{H_2(\Omega)}\, d\sigma\right) +$$

$$+ \|A_2^{\frac{1}{2}} u(\varepsilon)\|^2_{H_2(\Omega)},\quad 0 < \varepsilon \leq t < T(\varphi);$$

observe that any mapping w with

$$w' \in L^2((\varepsilon,t),H_2(\Omega)),$$
$$w(t) \in D(A_2) \text{ a.e. on } (\varepsilon,t),$$
$$A_2 w \in L^2((\varepsilon,t),H_2(\Omega)),$$
$$w'+A_2 w = F \in L^2((\varepsilon,t),H_2(\Omega))$$

has in fact the property that $w(t) \in D(A_2^{\frac{1}{2}})$ on $[\varepsilon,t]$ and

$$A_2^{\frac{1}{2}} w \in C^0([\varepsilon,t],H_2(\Omega)).$$

Namely, let $A_2 w \in C^0([\varepsilon,T],H_2(\Omega))$, $w \in C^1([\varepsilon,T],H_2(\Omega))$; then the inclusion in question follows by approximation from

$$\text{(IV.5.2)} \quad \int_\varepsilon^t 2\,\mathrm{Re}(A_2 w(t),w'(t))\,dt = \|A^{\frac{1}{2}} w(t)\|^2 - \|A^{\frac{1}{2}} w(\varepsilon)\|^2.$$

Now we want to estimate $\|M(u(\sigma))\|_{H_2(\Omega)}$. Let $s>2$. We have by Hölder's inequality

$$\|M(u(\sigma))\|^2_{H_2(\Omega)} = \|u(\sigma)\cdot\nabla(\sigma)\|^2_{(L^2(\Omega))^n}$$

$$\le c\|u(\sigma)\|^2_{(L^r(\Omega))^n}\|\nabla u(\sigma)\|^2_{(L^{\frac{2r}{r-2}}(\Omega))^{n^2}}$$

$$\le \|u(\sigma)\|^2_{(L^r(\Omega))^n}\|\nabla u(\sigma)\|^2_{(L^{\tilde{s}}(\Omega))^{n^2}}$$

with $\frac{1}{\tilde{s}} = \frac{1}{2}-\frac{1}{r}$. As in the proof of theorem IV.9.1 we get

$$\|\nabla u(\sigma)\|^2_{(L^s(\Omega))^{n^2}} \le \|u(\sigma)\|^{2\frac{n}{r}}_{H^{2,2}(\Omega)}\|\nabla u(\sigma)\|^{\frac{4}{s}}_{(L^2(\Omega))^{n^2}},$$

$$\le \|u(\sigma)\|^{2\frac{s-2}{s}}_{H^{2,2}(\Omega)}\|\nabla u(\sigma)\|^{\frac{4}{s}}_{(L^2(\Omega))^{n^2}}$$

and thus

$$\|M(u(\sigma))\|^2_{H_2(\Omega)} \leq c(\varepsilon)\|u(\sigma)\|^s_{(L^r(\Omega))^n}\|\nabla u(\sigma)\|^2_{(L^2(\Omega))^{n^2}} + \varepsilon\|u(\sigma)\|^2_{H^{2,2}(\Omega)}$$

for any $\varepsilon > 0$. This gives the a-priori estimate (cf. Gronwall's inequality)

$$\text{(IV.5.3)}\quad \|A_2^{\frac{1}{2}}u(t)\|^2_{H_2(\Omega)} + \int_\varepsilon^t \|A_2 u(\sigma)\|^2_{H_2(\Omega)}\, d\sigma \leq c(\|A_2^{\frac{1}{2}}u(\varepsilon)\|^2_{H_2(\Omega)} + \int_\varepsilon^{\tilde{T}} \|P_2 f(\sigma)\|^2_{H_2(\Omega)}\, d\sigma) \cdot \exp c \int_\varepsilon^t \|u(\sigma)\|^s_{(L^r(\Omega))^n}\, d\sigma$$

on any interval $[\varepsilon,\tilde{T}]$, $0 < \varepsilon < \tilde{T} < T(\varphi)$. Let us assume that $T(\varphi) < +\infty$. In view of (IV.5.2) and $w' \in L^2((0,T(\varphi)),(L^2(\Omega))^n)$ we have also proved that

$$\text{(IV.5.4)}\quad w_1(\delta) = \sup_{\substack{\varepsilon \leq t',t < T(\varphi)\\ |t'-t|\leq\delta}} \|A^{\frac{1}{2}}(u(t')-u(t))\|_{H_2(\Omega)} \to 0 \text{ for } \delta \to 0,$$

not making use of any restriction of the dimension n. Setting $n = 2,3,4$ we get by Sobolev that

$$w_2(\delta) = \sup_{\substack{0\leq t',t<T(\varphi)\\ |t'-t|<\delta}} \|u(t')-u(t)\|_{(L^n(\Omega))^n} \to 0$$

for $\delta \to 0$, i.e. u is uniformly continuous in the $(L^n(\Omega))^n$-norm with respect to t. In view of theorem II.3.6, (III.3.27), from this it follows that $T(\varphi) = +\infty$.

It remains to deal with the case $s = 2$. Then we get

$$\|M(u(\sigma))\|^2_{H^2(\Omega)} \leq \|u(\sigma)\|^2_{L^\infty(\Omega)} \|\nabla u(\sigma)\|^2_{(L^2(\Omega))^{n^2}}$$

and since $\|u(.)\|^2_{(L^\infty(\Omega))^n} \in \bigcap_{O<T} L^1((O,T))$ we can proceed as before. □

The preceding theorem is limited to $n = 3,4$. However, it may be also valid in the cases $n = 5,6$. This will be outlined in the following

<u>Remark:</u> Assuming that

$$\tilde{f} \in L^2((O,T),(H^{l+2,2}(\Omega))^n)$$

for some $l \in \mathbb{N}$ we first solve a linear problem: Let $\varepsilon > 0$, $\zeta(t) \equiv 0$ for $0 \leq t \leq \frac{\varepsilon}{2}$, $\zeta(t) \equiv 1$ for $t \geq \varepsilon$, $\zeta \in C^{0,1}([0,\infty))$. We consider

$$\text{(IV.5.5)} \quad (\zeta u)' - \nu\Delta\zeta u + \nabla\zeta\pi = \zeta f - \zeta' u, \qquad \nabla\cdot\zeta u = 0.$$

Since we must not care about compatibility conditions here we assume that we can get the a-priori estimate $(l = 2)$

$$\int_\varepsilon^t \|u''(\sigma)\|^2_{(L^2(\Omega))^n} d\sigma + \sup_{\varepsilon\leq\sigma\leq t} \|u'(\sigma)\|^2_{(H^{1,2}(\Omega))^n}$$

$$+ \int_\varepsilon^t \|u'(\sigma)\|^2_{(H^{1,2}(\Omega))^n} d\sigma + \int_\varepsilon^t \|u(\sigma)\|^2_{(H^{l+2,2}(\Omega))^n} +$$

$$+ \int_\varepsilon^t \|\nabla\pi(\sigma)\|^2_{(H^{1,2}(\Omega))^n} d\sigma + \sup_{\varepsilon\leq\sigma\leq t} \|u(\sigma)\|^2_{(H^{l+1,2}(\Omega))^2}$$

$$\leq c(\int_\varepsilon^t \|\tilde{f}(\sigma)\|^2_{(H^{1,2}(\Omega))^n} d\sigma + \int_\varepsilon^t \|\tilde{f}'(\sigma)\|^2_{(L^2(\Omega))^n} d\sigma +$$

$$+ \|u(\varepsilon)\|^2_{(H^{l+1,2}(\Omega))^2} + \int_\varepsilon^t \|u(\sigma)\|^2_{H^{1,2}(\Omega)} d\sigma +$$

$$+ \int_\varepsilon^t \|u'(\sigma)\|^2_{(L^2(\Omega))^n} d\sigma)$$

for any solution (u,π) with

$$u'' \in \bigcap_{0<\rho<T} L^2((\rho,T),(L^2(\Omega))^n)$$

$$u' \in \bigcap_{0<\rho<T} L^2((\rho,T),(H^{1,2}(\Omega))^n),$$

$$u \in \bigcap_{0<\rho<T} L^2((\rho,T),(H^{2+l,2}(\Omega))^n),$$

$$u' \in \bigcap_{0<\rho<T} C^0([\rho,T],(\overset{o}{H}{}^{1,2}(\Omega))^n,$$

$$u \in \bigcap_{0<\rho<T} C^0([\rho,T],(H^{1+l,2}(\Omega))^n) \cap$$

$$\cap \bigcap_{0<\rho<T} C^0([\rho,T],(\overset{o}{H}{}^{1,2}(\Omega))^n),$$

$$\pi' \in \bigcap_{0<\rho<T} L^2((\rho,T),(L^2(\Omega))^n),$$

$$\nabla\pi \in \bigcap_{0<\rho<T} L^2((\rho,T),(H^{l,2}(\Omega))^n)$$

(In fact this extension of theorem III.1.1 has been proved by Solonnikov [Sol2] for $n=3$; we assume, for a moment, it is valid also for $n \geq 4$). For $\tilde{f}$ we insert $-u\cdot\nabla u + f$, where f is from theorem IV.5.1 and we assume that u,π have all regularity properties being necessary to apply the preceding estimate; here u is the local strong solution from the proof of theorem IV.5.1 over some interval $[0,T]$, $T < T(\varphi)$, π is the pressure which belongs to u. We set $l = 2$. Of course

$$\int_\varepsilon^t \|u(\sigma)\|^2_{H^{1,2}} d\sigma$$

has already been estimated in any dimension (cf. the proof of theorem IV.5.1). Let us consider $(u\cdot\nabla u)'$ and assume that the chain rule holds. Then

$$(u\cdot\nabla u)' = u'\cdot\nabla u + u\cdot\nabla u'.$$

As in the proof of theorem IV.5.1 the term $u\cdot\nabla u'$ can be estimated to give

$$\|u\cdot\nabla u'(\sigma)\|^2_{(L^2(\Omega))^n} \le c\|u(\sigma)\|^2_{(L^r(\Omega))^n} \cdot\|u'(\sigma)\|^{\frac{2(s-2)}{2}}_{(H^{2,2}(\Omega))^n}\|u'(\sigma)\|^{\frac{4}{s}}_{(H^{1,2}(\Omega))^n}.$$

Moreover for $n=5,6$ by Sobolev

$$\begin{aligned}\|u'\cdot\nabla u(\sigma)\|^2_{(L^2(\Omega))^n} &\le c\|u'(\sigma)\|^2_{(L^6(\Omega))^n}\|\nabla u(\sigma)\|^2_{(L^3(\Omega))^{n'}}\\ &\le c\|u'(\sigma)\|^2_{(H^{2,2}(\Omega))^n}\|u(\sigma)\|^2_{(H^{2,2}(\Omega))^n}.\end{aligned}$$

Assuming again that we can apply the chain rule to get $\frac{\partial^2}{\partial x_i\partial x_j}u\cdot\nabla u$ we see that we have to estimate terms of the type $u\cdot\nabla\frac{\partial^2}{\partial x_i\partial x_j}u$ and $\frac{\partial u}{\partial x_i}\cdot\nabla\frac{\partial u}{\partial x_j}$. As in the proof of theorem IV.5.1 we get

$$\|u\cdot\nabla\frac{\partial^2}{\partial x_i\partial x_j}u(\sigma)\|^2_{(L^2(\Omega))^n} \le c\|u(\sigma)\|^2_{(L^r(\Omega))^n}\cdot \|u(\sigma)\|^{\frac{2(s-2)}{2}}_{(H^{4,2}(\Omega))^n}\|u(\sigma)\|^{\frac{4}{s}}_{(H^{3,2}(\Omega))^{n'}}$$

and again for $n=5,6$ by Sobolev

$$\begin{aligned}|\frac{\partial u}{\partial x_i}\cdot\nabla\frac{\partial u}{\partial x_j}(\sigma)\|^2_{(L^2(\Omega))^n} &\le c\|\nabla u(\sigma)\|^2_{(L^6(\Omega))^n}\|\nabla^2 u(\sigma)\|^2_{(L^3(\Omega))^{n'}}\\ &\le c\|u(\sigma)\|^2_{(H^{3,2}(\Omega))^n}.\end{aligned}$$

As before this gives a bound for $\sup_{\varepsilon\leq t\leq T} \|u(t)\|_{(H^{3,2}(\Omega))^n}$ and the relation $u \in C^0([\varepsilon,T(\varphi)],(H^{3,2}(\Omega))^n)$. As for

$$w_2(\delta) = \sup_{\substack{\varepsilon\leq t',t<T(\varphi)\\ |t'-t|\leq\delta}} \|u(t')-u(t)\|_{H^{3,2}(\Omega)},$$

this yields $w_2(\delta) \to 0$ for $\delta \to 0$. That in particular guarantees the uniform continuity of $u(.)$ over $[0,T(\varphi))$ with respect to the $L^n(\Omega)$-norm, $n=5,6$. We get the desired result $T(\varphi) = +\infty$ as before. □

Next we deal with the classes $C^0([0,T],(L^n(\Omega))^n)$ and $L^\infty((0,T),(L^n(\Omega))^n)$. As for the first class we have

Theorem IV.5.2: Let $\varphi \in H_n(\Omega)$. For some $p>n$ let $f \in C^{\frac{1}{2p}}([0,T],(L^p(\Omega))^n) \cap C^0((0,T],(H^{\frac{1}{p},p}(\Omega))^n)$ for all T, $T>0$. Let $\tilde{u}$ be a weak solution of the Navier-Stokes equations over $(0,T)\times\Omega$ for for every $T>0$ with boundary values $\tilde{u}(0)=\varphi$, $\tilde{u}|\partial\Omega=0$ in the sense of definition IV.1.1. Let

$$\tilde{u} \in C^0([0,T],(L^n(\Omega))^n),\ T>0.$$

Then

$$\tilde{u} \in C^0((0,T],(H^{2,p}(\Omega))^n)$$

for all $T>0$.

Proof: We take the strong solution u of the Navier-Stokes equations which has been constructed in theorem III.3.6. We know already that

$$\tilde{u}(t) = u(t) \text{ on } [0,T(\varphi))$$

(cf. theorem IV.4.4). We have to show that $T(\varphi) = +\infty$. From our assumptions it follows that

$$w_4(\delta) = \sup_{\substack{0 \le t',t < T(\varphi),\\ |t'-t| \le \delta}} \|\tilde{u}(t')-\tilde{u}(t)\|_{(L^n(\Omega))^n}$$

$$\le \sup_{\substack{0 \le t',t \le T(\varphi),\\ |t'-t| \le \delta}} \|u(t')-u(t)\|_{(L^n(\Omega))^n}$$

$$\to 0 \text{ for } \delta \to 0.$$

Thus theorem IV.3.6, (III.3.27) completes the proof. □

In view of the fact that $L^\infty((0,T),(L^n(\Omega))^n)$ is a uniqueness class for weak solutions, provided the data φ, f fulfill suitable regularity assumptions, the following theorem may appear rather natural; but so far it was not possible to verify its assumptions unless in a few special cases (cf. [W10] for n=2, [W4] for other cases). The theorem in question says that stability of weak solutions (or continuous dependence on the data) in the norm of $L^\infty((0,T),(L^n(\Omega))^n)$ is sufficient for regularity.

<u>Theorem IV.5.3:</u> <u>Let</u> $\varphi^1,\varphi^2 \in (H^{2,p}(\Omega))^n \cap (\overset{o}{H}{}^{1,p}(\Omega))^n \cap H_p(\Omega)$ <u>for some</u> $p > n$, <u>let</u>

$$f^1,f^2 \in C^{\frac{1}{2p}}([0,T],(L^p(\Omega))^n) \cap C^0((0,T],(H^{\frac{1}{p},p}(\Omega))^n)$$

<u>for all</u> $T > 0$.

<u>If</u> u^1,u^2 <u>are weak solutions of the Navier-Stokes equations in the sense of definition</u> IV.1.1 <u>over</u> $(t_o,T)\times\Omega$ <u>for an</u> $t_o > 0$ <u>and and every</u> $T > t_o$ <u>with boundary values</u> $u^1(t_o) = \varphi^1$, $u^2(t_o) = \varphi^2$, $u^1|\partial\Omega = 0$, $u^2|\partial\Omega = 0$ <u>and inhomogeneous terms</u> f^1,f^2, <u>which moreover fulfill</u>

$$u^1 \in L^\infty((t_o,T),(L^n(\Omega))^n),$$
$$u^2 \in L^\infty((t_o,T),(L^n(\Omega))^n),\ T > 0,$$

then we assume that for any $t_o > 0$ there exists a function g_{t_o} with

$$\operatorname*{ess\,sup}_{t_o<t<T} \|u^1(t)-u^2(t)\|^n_{(L^n(\Omega))^n}$$

$$\leq g_{t_o}(T, \|\varphi^1-\varphi^2\|_{(H^{2,p}(\Omega))^n}, \operatorname*{ess\,sup}_{t_o<t<T} \|f^1(t)-f^2(t)\|_{(L^p(\Omega))^n}),$$

$$T > t_o;$$

here the function $g_{t_o}: \mathbb{R}^+ \times \mathbb{R}^+ \times \mathbb{R}^+ \to \mathbb{R}^+$ is not depending on u^1, u^2 and has the property that

$$g_{t_o}(T, r_1, r_2) \to 0, \quad T > 0,$$

if $r_1, r_2 \to 0$. Under the assumption above the following conclusion is valid: Let u be any weak solution over $(0,T) \times \Omega$ for all $T > 0$ with boundary values $u(0) = \varphi$, $u|\partial\Omega = 0$ and inhomogeneous term f where

$$\varphi \in H_n(\Omega),$$

$$f \in C^{\frac{1}{2p}}([0,T], (L^p(\Omega))^n) \cap C^o((0,T], (H^{\frac{1}{p},p}(\Omega))^n) \text{ for } T > 0;$$

moreover

$$u \in L^\infty((0,T), (L^n(\Omega))^n).$$

Then

$$u \in C^o((0,T], (H^{2,p}(\Omega))^n) \text{ for all } T > 0.$$

Remark: 1. For the proof we need Solonnikov's potential theoretical estimates in any number of space dimensions (theorem III.1.1, where n was equal to 2 or 3), but in a more restricted form, namely for initial values from $H^{2,p}(\Omega) \cap \overset{o}{H}{}^{1,p}(\Omega) \cap H_p(\Omega)$; the proof of this, following the lines of Solonnikov's paper, has been outlined in [W 2], [W 4] (cf. the discussion at the end of III.4). Speaking strictly the preceding theorem is thus limited to $n = 2,3$.

2. It may be pointed out that theorems IV.5.1, IV.5.2, IV.5.3 remain still valid if the occurring data φ, f fulfill

$$\varphi \in H_n(\Omega),$$
$$f \in C^{\frac{1}{2p}}([0,T],(L^p(\Omega))^n).$$

In some case the condition on f may be replaced by $f \in C^0([0,T], (L^n(\Omega))^n) \cap C^{\frac{1}{2p}}((0,T],(L^p(\Omega))^n)$ but we do not want to emphasize this point.

<u>Proof of theorem IV.5.3:</u> Let $4n > p > n$. Let us consider the equations

$$\begin{aligned} u_\sigma' - \nu\Delta u_\sigma + u_\sigma \cdot \nabla u_\sigma + \nabla\pi_\sigma &= \sigma f \\ \nabla \cdot u_\sigma &= 0, \\ u_\sigma(0) &= \sigma\varphi, \\ u_\sigma(t)\,|\,\partial\Omega &= 0, \quad 0 \leq t \leq T, \quad 0 \leq \sigma \leq 1 \end{aligned}$$

within the class of admissible solutions

$$\nabla\pi_\sigma \in \bigcap_{0<\varepsilon<T} L^p((\varepsilon,T),(L^p(\Omega))^n),$$

$$u_\sigma \in C^0([0,T],(L^n(\Omega))^n),$$

$$u_\sigma' \in \bigcap_{0<\varepsilon<T} L^p((\varepsilon,T),(L^p(\Omega))^n),$$

$$u_\sigma \in \bigcap_{0<\varepsilon<T} L^p((\varepsilon,T),(H^{2,p}(\Omega))^n \cap (\overset{\circ}{H}{}^{1,p}(\Omega))^n)$$

$$u_\sigma \in \bigcap_{0<\varepsilon<T} C^0([\varepsilon,T],(W^{2-2/p,p}(\Omega))^n)$$

$$\nabla\cdot u_\sigma(t) = 0 \text{ on } (0,T],$$

u_σ is a weak solution of the Navier-Stokes equations over $(0,T)\times\Omega$ with boundary values $u_\sigma(0) = \sigma\varphi$, $u_\sigma|\partial\Omega = 0$ and inhomogeneous term σf in the sense of definition IV.1.1,

$$u_\sigma \in L^\infty((0,T),(L^n(\Omega))^n),\ 0 \leq \sigma \leq 1,\ T > 0.$$

For $\sigma = 0$ we have as unique solution $u_0 \equiv 0$. According to theorem IV.4.1 each solution u_σ is determined uniquely. According to theorems IV.4.5, III.3.6 each solution u_σ can be reconstructed on some interval $[0,T^*(\sigma)]$, $T^*(\sigma) > 0$. We want to show that we can choose for $T^*(\sigma)$ one and the same positive quantity $T^* > 0$, $0 \leq \sigma \leq 1$. For this we go back to the proofs of theorem III.3.6, II.3.8. When we have constructed there the local strong solution on a first interval of existence $[0,\tilde{T}]$, the quantity $\tilde{T} > 0$ has depended only on the smallness of

$$\| t^{\frac{1}{4}} A_n^{\frac{1}{4}} e^{-tA_n}\varphi\|_{H^n(\Omega)}.$$

In our case this means that $T^*(\sigma)$ depends on the smallness of

$$\| t^{\frac{1}{4}} A_n^{\frac{1}{4}} e^{-tA_n}\sigma\varphi\|_{H_n(\Omega)} \leq \| t^{\frac{1}{4}} A_n^{\frac{1}{4}} e^{-tA_n}\varphi\|_{H_n(\Omega)}.$$

Thus we can in fact choose for any $T^*(\sigma)$ one and the same $T^* > 0$. It is also clear from the proofs of theorems III.3.6, II.3.8 that on $[0,T^*]$

$$\| t^{\frac{1}{4}} A_n^{\frac{1}{4}} u_\sigma(\tilde{\sigma})\|_{H_n(\Omega)} \leq M(t),\ 0 \leq \tilde{\sigma} \leq t,\ 0 \leq \sigma \leq 1,$$

with $M: \mathbb{R}^+ \to \mathbb{R}^+$ not depending on σ and $M(t) \to 0$ for $t \to 0$. Now let $\eta > 0$. Then again from the proofs of theorems III.3.6, II.3.8 it follows that for $0 < \varepsilon < \frac{1}{2}$, $0 < t \leq T^*$ we get

$$t^{\frac{1}{2}-\varepsilon}\|A_n^{\frac{1}{2}-\varepsilon}u_{\sigma_2}(t)-A_n^{\frac{1}{2}-\varepsilon}u_{\sigma_1}(t)\|_{H_n(\Omega)}$$

$$\leq c|\sigma_2-\sigma_1|\|\varphi\|_{H_n(\Omega)}+c(f,T)|\sigma_2-\sigma_1|+$$

$$+ct^{\frac{1}{2}-\varepsilon}\int_0^t\frac{1}{(t-\tilde\sigma)^{1-\varepsilon}}\frac{M(\sigma)}{\tilde\sigma^{1/2}}\tilde\sigma^{\frac{1}{4}}\|A_n^{\frac{1}{4}}(u_{\sigma_2}(\tilde\sigma)-u_{\sigma_1}(\tilde\sigma))\|_{H_n(\Omega)}\,d\tilde\sigma,$$

$$\leq c|\sigma_2-\sigma_1|\|\varphi\|_{H_n(\Omega)}+c(f,T)|\sigma_2-\sigma_1|+$$

$$+ct^{\frac{1}{2}-\varepsilon}\int_0^t\frac{1}{(t-\tilde\sigma)^{1-\varepsilon}}\frac{M(t)}{\tilde\sigma^{\frac{3}{4}-\varepsilon}}\tilde\sigma^{\frac{1}{2}-\varepsilon}\|A_n^{\frac{1}{4}}(u_{\sigma_2}(\tilde\sigma)-u_{\sigma_1}(\tilde\sigma))\|_{H_n(\Omega)}\,d\sigma,$$

$$\leq c|\sigma_2-\sigma_1|\|\varphi\|_{H_n(\Omega)}+c(f,T)|\sigma_2-\sigma_1|+$$

$$+\frac{cM(t)}{t^{\frac{3}{4}-2\varepsilon}}t^{\frac{1}{2}-\varepsilon}\sup_{0<\tilde\sigma\leq t}\tilde\sigma^{\frac{1}{2}-\varepsilon}\|A_n^{\frac{1}{4}}(u_{\sigma_2}(\tilde\sigma)-u_{\sigma_1}(\tilde\sigma))\|_{H_n(\Omega)}.$$

Taking T* sufficiently small and taking consecutively $\varepsilon=\frac{1}{4}$, $\varepsilon=\frac{1}{8}$ we thus arrive at

$$t^{\frac{1}{4}}\|A_n^{\frac{1}{4}}(u_{\sigma_2}(t)-u_{\sigma_1}(t))\|_{H_n(\Omega)}$$

$$\leq c|\sigma_2-\sigma_1|\|\varphi\|_{H_n(\Omega)}+c(f,T)|\sigma_2-\sigma_1|,$$

$$\|A_n^{\frac{3}{8}}(u_{\sigma_2}(t)-u_{\sigma_1}(t))\|_{H_n(\Omega)}\leq\frac{c}{t^{\frac{3}{8}}}|\sigma_2-\sigma_1|(\|\varphi\|_{H_n(\Omega)}+c(f,T)).$$

By theorems III.2.3, III.2.4 we have

$$D(A_n^{\frac{3}{8}})\subset D(A_p^{\varepsilon'})$$

for some ε', $0 < \varepsilon' < \frac{1}{2p}$; here we need $n < p < 4n$. Thus

$$\|A_p^{\varepsilon'}(u_{\sigma_2}(t)-u_{\sigma_1}(t))\|_{H_p(\Omega)}$$

$$\leq \frac{c}{t^{\frac{3}{8}}}|\sigma_2-\sigma_1|(\|\varphi\|_{H_n(\Omega)}+c(f,T)).$$

Observe that T* does not depend on σ_2,σ_1. Taking the integral equations in $H_p(\Omega)$ for u_{σ_2}, u_{σ_1} we get with the aid of the proof of proposition III.3.1

$$\text{(IV.5.5)}\quad \|A_p(u_{\sigma_2}(t)-u_{\sigma_1}(t))\|_{H_p(\Omega)} \leq \frac{c}{(t-\eta)^{1-\varepsilon'}}\cdot$$

$$\cdot\frac{1}{\eta^{\frac{3}{8}}}|\sigma_2-\sigma_1|\,\|\varphi\|_{H_n(\Omega)} + c(T,f)|\sigma_2-\sigma_1|$$

$$+\int_\eta^t \frac{c}{(t-\tilde\sigma)^{1-\varepsilon'}}\|A_p^{\varepsilon'}(M(u_{\sigma_2}(\tilde\sigma))-M(u_{\sigma_1}(\tilde\sigma)))\|_{H_p(\Omega)}\,d\tilde\sigma,$$

$$\leq \frac{c}{(t-\eta)^{1-\varepsilon'}}\,\frac{1}{\eta^{\frac{3}{8}}}|\sigma_2-\sigma_1|\,\|\varphi\|_{H_n(\Omega)} + c(T,f)|\sigma_2-\sigma_1| +$$

$$+\int_\eta^t \frac{c}{(t-\eta)^{1-\varepsilon'}}(\|A_p^{1-\varepsilon''}u_{\sigma_2}(\tilde\sigma)\|_{H_p(\Omega)}+\|A_p^{1-\varepsilon''}u_{\sigma_1}(\tilde\sigma)\|_{H_p(\Omega)})\cdot$$

$$\cdot\|A_p^{1-\varepsilon''}(u_{\sigma_2}(\tilde\sigma)-u_{\sigma_1}(\tilde\sigma))\|_{H_p(\Omega)}\,d\tilde\sigma$$

for some $\eta \in (0,T^*)$ and some ε'', $0<\varepsilon''<\varepsilon'$, ρ_1, where ρ_1 is the quantity from theorem III.3.3. [1] Taking $\frac{\eta}{2}$ instead of η we get

---

[1] In the same way as we did for (III.3.20) one can prove an analogous estimate for $\|A^{\frac{\hat\rho}{2}}P_p(u-v)\cdot\nabla u\|_{H_p(\Omega)}$ and $\|A^{\frac{\hat\rho}{2}}P_p v\cdot\nabla(u-v)\|_{H_p(\Omega)}$.

$$\|A_p^{1-\varepsilon''}u_{\sigma_i}(t)\|_{H_p(\Omega)} \le \frac{c}{(t-\frac{\eta}{2})^{1-\varepsilon''}}\frac{\|\varphi\|_{H^n(\Omega)}}{(\frac{\eta}{2})^{3/8}} + c(T,f)$$

$$+ \int_{\frac{\eta}{2}}^{t} \frac{c}{(t-\tilde{\sigma})^{1-\varepsilon''}}\|A_p^{1-\varepsilon''}u_{\sigma_i}(\tilde{\sigma})\|_{H_p(\Omega)}\|u_{\sigma_i}(\tilde{\sigma})\|_{H_p(\Omega)}\, d\sigma,$$

$$\le \frac{c}{(t-\frac{\eta}{2})^{1-\varepsilon''}}\frac{\|\varphi\|_{H_n(\Omega)}}{(\frac{\eta}{2})^{3/8}} + c(T,f)$$

$$+ \int_{\frac{\eta}{2}}^{t} \frac{c}{(t-\tilde{\sigma})^{1-\varepsilon''}}\|A_p^{1-\varepsilon''}u_{\sigma_i}(\tilde{\sigma})\|\frac{\|\varphi\|_{H^n(\Omega)}}{(\frac{\eta}{2})^{3/8}}\, d\tilde{\sigma},\quad i=1,2.$$

Thus as on pp. 5,6 we get

$$\|A_p^{1-\varepsilon''}u_{\sigma_i}(t)\|_{H_p(\Omega)} \le \frac{c}{(t-\frac{\eta}{2})^{1-\varepsilon''}}\left(\frac{\|\varphi\|_{H_n(\Omega)}}{(\frac{\eta}{2})^{3/8}} + c(T,f)\right)\cdot$$

$$\cdot\exp\left(\frac{c\|\varphi\|_{H_n(\Omega)}}{(\frac{\eta}{2})^{3/8}}t\right),$$

$$\|A_p^{1-\varepsilon''}u_{\sigma_i}(t)\|_{H_p(\Omega)} \le \frac{c}{\eta^{1-\varepsilon''+3/8}}(\|\varphi\|_{H_n(\Omega)}+c(T,f))$$

$$\cdot\exp\left(\frac{c\|\varphi\|_{H_n(\Omega)}}{\eta^{3/8}}T^*\right),\quad \eta\le t\le T^*.$$

Inserting this estimate into (IV.5.5) we arrive at

$$\text{(IV.5.6)}\quad \|A_p(u_{\sigma_2}(t)-u_{\sigma_1}(t))\|_{H_p(\Omega)} \le \frac{c(\eta,T^*,\|\varphi\|_{H_n(\Omega)})}{(t-\eta)^{1-\varepsilon'}}$$

$$\cdot(|\sigma_2-\sigma_1|\ \|\varphi\|_{H_n(\Omega)} + |\sigma_2-\sigma_1|c(T,f)),$$

where the constants do not depend on σ_2,σ_1. Now we want to fix an η, $0<2\eta<T^*$. We want to consider the solutions u_σ on $[2\eta,T]$. For the difference $u_{\sigma_2}-u_{\sigma_1}$ we have

$$(u_{\sigma_2}-u_{\sigma_1})' - \nu\Delta(u_{\sigma_2}-u_{\sigma_1}) + \nabla(\pi_{\sigma_2}-\pi_{\sigma_1})$$

$$= (\sigma_2-\sigma_1)f - u_{\sigma_2}\cdot\nabla u_{\sigma_2} + u_{\sigma_1}\cdot\nabla u_{\sigma_1}.$$

Theorem III.1.1 for arbitrary n and proposition IV.2.2 then give

$$\int_{2\eta}^{\tilde{T}} \|u'_{\sigma_2}(\tilde{\sigma})-u'_{\sigma_1}(\tilde{\sigma})\|^p_{(L^p(\Omega))^n}\, d\tilde{\sigma} +$$

$$+ \int_{2\eta}^{\tilde{T}} \|u_{\sigma_2}(\tilde{\sigma})-u_{\sigma_1}(\tilde{\sigma})\|^p_{(H^{2,p}(\Omega))^n}\, d\tilde{\sigma} +$$

$$+ \sup_{2\eta\leq\tilde{\sigma}\leq\tilde{T}} \|u_{\sigma_2}(\tilde{\sigma})-u_{\sigma_1}(\tilde{\sigma})\|^p_{(W^{2-2/p,p}(\Omega))^n}$$

$$\leq c(\int_{2\eta}^{\tilde{T}} |\sigma_2-\sigma_1|^p \|f(\tilde{\sigma})\|^p_{(L^p(\Omega))^n}\, d\tilde{\sigma} + \|A_p(u_{\sigma_2}(2\eta)-u_{\sigma_1}(2\eta))\|^p_{H_p(\Omega)}$$

$$+ \int_{2\eta}^{\tilde{T}} [g^{2p}_{2\eta}(T, \|A_p(u_{\sigma_2}(2\eta)-u_{\sigma_1}(2\eta))\|_{H_p(\Omega)},$$

$$\operatorname{ess\,sup}_{2\eta\leq\tilde{\sigma}\leq\tilde{T}} \|(\sigma_2-\sigma_1)f(\tilde{\sigma})\|_{L^p(\Omega)})\|u_{\sigma_2}(\tilde{\sigma})-u_{\sigma_1}(\tilde{\sigma})\|^p_{(H^2,p(\Omega))^n} +$$

$$g^{\frac{p}{2}}_{2\eta}(T, \|A_p(u_{\sigma_2}(2\eta)-u_{\sigma_1}(2\eta))\|_{H_p(\Omega)},$$

$$\operatorname{ess\,sup}_{2\eta\leq\tilde{\sigma}\leq\tilde{T}} \|(\sigma_2-\sigma_1)f(\tilde{\sigma})\|_{L^p(\Omega)})\|u_{\sigma_2}(\tilde{\sigma})-u_{\sigma_1}(\tilde{\sigma})\|^p_{(H^2,p(\Omega))^n}]\, d\sigma$$

$$+ c\int_{2\eta}^{\tilde{T}} (\|u_{\sigma_1}(\tilde{\sigma})\|^{3p}_{(L^{3p}(\Omega))^n} + \|\nabla u_{\sigma_1}(\tilde{\sigma})\|^{\frac{3}{2}p}_{(L^{3p/2}(\Omega))^{n^2}} +$$

$$+ \|u_{\sigma_1}(\tilde{\sigma})\cdot\nabla u_{\sigma_1}(\tilde{\sigma})\|^p_{(L^p(\Omega))^n}\, d\tilde{\sigma}),\quad 2\eta \leq \tilde{T} \leq T;$$

in order to derive this inequality we have used the trivial estimates $\|u_{\sigma_2}\|^{3p}_{L^{3p}(\Omega)} \leq c(\|u_{\sigma_2}-u_{\sigma_1}\|^{3p}_{(L^{3p}(\Omega))^n}+\|u_{\sigma_1}\|^{3p}_{(L^{3p}(\Omega))^n})$,

$$\|\nabla u_{\sigma_2}\|^{\frac{3}{2}p}_{(L^{3p/2}(\Omega))^{n^2}} \le c(\|\nabla u_{\sigma_2}-\nabla u_{\sigma_1}\|^{\frac{3}{2}p}_{(L^{3p/2}(\Omega))^{n^2}} +$$

$+ \|\nabla u_{\sigma_1}\|^{\frac{3}{2}p}_{(L^{3p/2}(\Omega))^{n^2}})$ and the second inequality in proposition IV.2.2. Taking $|\sigma_2-\sigma_1|$ small enough (in dependence of η) and using (IV.5.6) we get an a-priori estimate for

$$|||u_{\sigma_2}-u_{\sigma_1}|||^p_{(2\eta,\tilde{T})p} = \int_{2\eta}^{\tilde{T}} \|u'_{\sigma_2}(\tilde{\sigma})-u'_{\sigma_1}(\tilde{\sigma})\|^p_{(L^p(\Omega))^n}\, d\tilde{\sigma} +$$

$$+ \int_{2\eta}^{\tilde{T}} \|u_{\sigma_2}(\tilde{\sigma})-u_{\sigma_1}(\tilde{\sigma})\|^p_{(H^{2,p}(\Omega))^n}\, d\tilde{\sigma} +$$

$$+ \sup_{2\eta\leq\tilde{\sigma}\leq\tilde{T}} \|u_{\sigma_2}(\tilde{\sigma})-u_{\sigma_1}(\tilde{\sigma})\|^p_{(W^{2-2/p,p}(\Omega))^n}$$

in terms of the data and of

$$\int_{2\eta}^{\tilde{T}} (\|u_{\sigma_1}(\tilde{\sigma})\|^{3p}_{(L^{3p}(\Omega))^n} + \|\nabla u_{\sigma_1}(\tilde{\sigma})\|^{\frac{3}{2}p}_{(L^{3p/2}(\Omega))^{n^2}})\, d\tilde{\sigma}$$

$$+ \int_{2\eta}^{\tilde{T}} \|u_{\sigma_1}(\tilde{\sigma})\cdot\nabla u_{\sigma_1}(\tilde{\sigma})\|^p_{(L^p(\Omega))^n}\, d\tilde{\sigma}$$

$$\leq c \int_{2\eta}^{\tilde{T}} \|u_{\sigma_1}(\tilde{\sigma})\|^p_{(H^{2,p}(\Omega))^n}(\|u_{\sigma_1}(\tilde{\sigma})\|^{2p}_{(L^n(\Omega))^n}+\|u_{\sigma_1}(\tilde{\sigma})\|^{\frac{1}{2}p}_{(L^n(\Omega))^n})\,d\tilde{\sigma}$$

(cf. proposition IV.2.2); since by Sobolev

$$\text{(IV.5.7)}\quad \|u\|_{(C^0(\bar{\Omega}))^n} \leq c\|u\|_{(W^{2-2/p,p}(\Omega))^n},\quad u \in (W^{2-2/p,p}(\Omega))^n$$

we have estimated $|||u_{\sigma_2}-u_{\sigma_1}|||^p_{(2\eta,\tilde{T}),p}$ in terms of $|||u_{\sigma_1}|||^p_{(2\eta,\tilde{T}),p}$. Since, however, the smallness of $|\sigma_2-\sigma_1|$ did not depend on u_{σ_1} we can exhaust stepwise the σ-interval [0,1] in order to get an a-priori estimate for $|||u_1|||^p_{(2n,\tilde{T}),p}$. Since $u_1 = u$ as long as u exists as a strong solution, in view of (IV.5.7) and theorem III.3.4 we have shown that $T(\varphi) = +\infty$, where $T(\varphi)$ is the length of the maximal interval of existence of u according to theorem III.3.6. Of course any $T(\sigma\varphi)$, $0 \leq \sigma \leq 1$, is also equal to $+\infty$. In the case $p \geq 4n$ we can deal first with exponents $\tilde{p}$, $n < \tilde{p} < 4n$; this gives the a-priori estimates being necessary. Our theorem is proved. □

Next we deal with possible singularities of weak solutions. The theorem to follow was proved in [SOW1].

<u>Theorem IV.5.4:</u> <u>Let</u> $\varphi \in H_n(\Omega)$, <u>let</u> $f \in C^{\frac{1}{2p}}([0,T],(L^p(\Omega))^n) \cap C^0((0,T],(H^{\frac{1}{p},p}(\Omega))^n)$ <u>for some</u> $p > n$ <u>and all</u> $T > 0$. <u>Let</u> u <u>be a weak solution of the Navier-Stokes equations over</u> $(0,T)\times\Omega$ <u>for every</u> $T > 0$ <u>with boundary values</u> $u(0) = \varphi$, $u|\partial\Omega = 0$ <u>and inhomogeneous term</u> f. <u>Let</u>

$$u \in L^\infty((0,T),(L^n(\Omega))^n) \text{ for all } T > 0.$$

<u>Then</u>

$$\mathbb{R}^+ = [0,+\infty) = \bigcup_{\nu=1}^{\infty} J_\nu \cup S$$

<u>where the</u> J_ν <u>are open intervals with</u>

$$u \in C^0(J_\nu,(H^{2,p}(\Omega))^n),$$

$$J_\nu \cap J_\mu = \emptyset, \quad \nu \neq \mu,$$

<u>and where</u>

$$S = [0,+\infty) - \bigcup_{\nu=1}^{\infty} J_\nu$$

<u>contains at most countably many points</u> t. <u>The points of</u> S <u>are exactly the points</u> 0 <u>and all points</u> t, <u>where</u> u <u>is continuous</u>

from the right with respect to t in the $H_n(\Omega)$-norm but discontinuous from the left. u is thus continuous from the right at any $t \geq 0$. If

(IV.5.8) $\|P_n f(t)\|_{H_n(\Omega)} + \|u(T_1)\|_{H_n} \leq \varepsilon_1$

for all $t \geq T_1$ for some $T_1 \geq 0$ where ε_1 is a certain positive number depending on A_n, then

$$u \in C^0([T_1+\infty),(H^{2,p}(\Omega))^n).$$

Thus a weak solution from $L^\infty((0,T),L^n(\Omega))$ has at most countably many singularities.

Proof of theorem IV.5.4: We know already from the proof of theorem IV.4.2 that

$$u(t) \in H_n(\Omega),$$

$$u:[0,+\infty) \to H_n(\Omega) \text{ is weakly continuous.}$$

Moreover we have already proved that $L^\infty((t_0,\tilde{T}),(L^n(\Omega))^n)$ is a uniqueness class for weak solutions over $(t_0,\tilde{T})\times\Omega$ (this is just a consequence from theorem IV.2.2 as we have remarked right after the proof of theorem IV.2.2); thus u can be reconstructed as a strong solution in the sense of theorem III.3.6 on some interval $[t,t+\varepsilon(t))$ for any $t \geq 0$; here $\varepsilon(t)$ is simply the number $T(u(t))$ depending on $u(t)$ which has been given in theorem III.3.6. For an arbitrary fixed $T > 0$ any finite sum

$$\sum_{i=1}^{N} l(t_i) \text{ is } \leq T,$$

where $t_i \in [0,T)$, $l(t_i)$ = length of $[t_i,t_i+T(u(t_i))) \cap [0,T]$, $[u(t_i),u(t_i)+T(u(t_i))) \cap [u(t_k),u(t_k)+T(u(t_k))) = \phi$, $i \neq k$, $1 \leq i,k \leq N$; we get that there exist at most countably many pairwise disjoint intervals $[t,t+T(u(t)))$ as described above in

$[0,\infty)$; we consider those where $t_\nu = 0$ or $t_\nu > 0$ and u is discontinuous from the left in t_ν; in what follows they are also denoted by $[t_1,t_1+T(u(t_1)))$, $[t_2,t_2+T(u(t_2)))$,...; we set

$$J_\nu = (t_\nu,t_\nu+T(u(t_\nu)));$$

the $[t_\nu,t_\nu+T(u(t_\nu)))$ as just introduced are simply the components of connectedness of

$$\bigcup_{t\geq 0} [t,t+T(u(t)))$$

and therefore

$$\bigcup_{t\geq 0} [t,t+T(u(t))) = \bigcup_{\nu=1}^{\infty} [t_\nu,t_\nu+T(u(t_\nu))),$$

namely: Assume that there is a $\tilde{t} \in \bigcup_{t\geq 0} [t,t+T(u(t)))$ which is not contained in $\bigcup_{i=1}^{\infty} [t_\nu,t_\nu+T(u(t_\nu)))$; then consider the interval $[\tilde{t},\tilde{t}+T(u(\tilde{t})))$. If it has a point in common with some $[t_\nu,t_\nu+T(u(t_\nu)))$ then either $\tilde{t} \in [t_\nu,t_\nu+T(u(t_\nu)))$ or $t_\nu \in [\tilde{t},\tilde{t}+T(u(\tilde{t})))$ and u is continuous in t_ν; in any case we get a contradiction. We set

$$S = [0,+\infty) - \bigcup_{\nu=1}^{\infty} J_\nu .$$

Now let $t \in S$. We can reconstruct u as a local strong solution on $[t,t+T(u(t)))$ with initial value $u(t)$. Thus $[t,t+T(u(t)))$ is contained in one of the $[t_\nu,t_\nu+T(u(t_\nu)))$, but since $t \notin J_\nu$ we only have the possibilities

$$t = t_\nu \text{ or,}$$

$$t = t_\nu+T(u(t_\nu)).$$

Thus S is countable. In the points t_ν or $t_\nu+T(u(t_\nu))$ u is continuous from the right in $H_n(\Omega)$-norm with respect to t, since u can be reconstructed as a local strong solution on $[t_\nu,t_\nu+T(u(t_\nu)))$. Of course u is continuous in J_ν in the $H_n(\Omega)$-norm for the same reason. If $t_\nu \in S$, $t_\nu > 0$, then u is discontinuous from the left in the $H_n(\Omega)$-norm in $t_\nu + T(u(t_\nu))$: Namely if we assume that u is in $C^o([t_\nu, t_\nu + T(u(t_\nu)]\ , H_n(\Omega))$ then according to theorem III.3.6 u can be continued as a local strong solution on $[t_\nu, t_\nu + T(u(t_\nu)) + \varepsilon]$ for some $\varepsilon > 0$; this being a contradiction we have proved the assertion. Since for each ν there is at most one and only one μ with

$$t_\nu + T((u(t_\nu))) = t_\mu$$

the points $t > 0$ of S consist exactly of the points $t_\nu + T(u(t_\nu))$. So far we have proved the first part of our theorem. The second part may be the easy one. If (IV.5.8) is fulfilled then we can construct a strong $\tilde{u}$ solution on $[T_1, +\infty)$ with initial value $\tilde{u}(T_1)$, homogeneous term f, and with $\tilde{u} \in C^o((T_1, +\infty), (H^{2,p}(\Omega))^n)$; this can be seen directly from theorem III.4.1, part 4. According to what have remarked after theorem IV.4.2 (and frequently used) $\tilde{u}$ and u coincide on $[T_1, +\infty)$. Our theorem is proved. □

Now we want to give a generalization of theorem IV.5.4 which turns out to be useful for dimensions $n = 3,4$.

<u>Theorem IV.5.5:</u> 1. <u>Let</u> $\varphi \in H_2(\Omega)$, <u>let</u> $f \in C^{\frac{1}{2p}}([0,T], (L^p(\Omega))^n) \cap C^o((0,T], (H^{\frac{1}{p},p}(\Omega))^n)$ <u>for some</u> $p > n$ <u>and for all</u> $T > 0$. <u>Let</u> u <u>be a weak solution of the Navier-Stokes equations over</u> $(0,T) \times \Omega$ <u>for every</u> $T > 0$ <u>with boundary values</u> $u(0) = \varphi$, $u | \partial\Omega = 0$ <u>and inhomogeneous term</u> f. <u>Let</u>

$$u(t) \in (L^n(\Omega))^n \ \underline{\text{a.e. on}}\ (0, +\infty),$$

<u>let the energy inequality</u>

$$\text{(IV.5.9)}\quad \|u(t)\|^2_{(L^2(\Omega))^n} + 2\nu \int_s^t \|\nabla u(\sigma)\|^2_{(L^2(\Omega))^{n^2}} d\sigma$$

$$\leq \|u(s)\|^2_{(L^2(\Omega))^n} + 2 \int_s^t (f(\sigma), u(\sigma))\, d\sigma$$

be fulfilled for almost all $s \in (0,+\infty)$ and all t, $t \geq s$. Then

(IV.5.10) $$(0,+\infty) = \bigcup_{\nu=1}^{\infty} J_\nu \cup S$$

where the J_ν are open intervals with

$$u \in C^0(J_\nu, (H^{2,p}(\Omega))^n),$$

$$J_\nu \cap J_\mu \neq 0, \ \nu \neq \mu,$$

and where

$$S \cup \{0\} = [0,+\infty) - \bigcup_{\nu=1}^{\infty} J_\nu$$

is a closed set of measure 0. If

(IV.5.11) $$\|P_n f(t)\|_{H_n(\Omega)} + \|u(T_1)\|_{H_n(\Omega)} \leq \varepsilon_1$$

for all $t \geq T_1$ for some $T_1 \geq 0$, where ε_1 is a certain positive number depending on A_n, then

$$u \in C^0([T_1,+\infty), (H^{2,p}(\Omega))^n).$$

2. Let $n = 3$ or $= 4$. Let φ, f be as above. Let u be a weak solution as above with the exception that only (IV.5.9) is assumed to be valid. Then the conclusion (IV.5.10) holds for u. If additionally

$$f \in L^2((0,+\infty), (H^{-1,2}(\Omega))^n)$$

and if

(IV.5.12) $$\|P_n f(t)\|_{H_n(\Omega)} \leq \varepsilon_1$$

for all $t \geq \tilde{T}_1$ for some $\tilde{T}_1 \geq 0$, where ε_1 is a certain positive number depending on A_n, then

$$u \in C^0([T_1,+\infty), (H^{2,p}(\Omega))^n)$$

for some $T_1 \geq 0$.

Remark: 1. In part 2. of the preceding theorem the assumption: $u(t) \in (L^n(\Omega))^n$ of part 1. has been dropped. Theorem IV.5.5 can be considered as a generalization of Leray's celebrated structure theorem. Leray has proved the second part of the present theorem in the case $n = 3$, $\Omega = \mathbb{R}^3$; moreover he shows that

$$\sum_{\nu=1}^{\infty} |J_\nu|^{\frac{1}{2}} < +\infty$$

([Ler]). His result was carried over to the case $n = 4$ by Kato [K]. In the case $n = 2$ any weak solution is regular for $t > 0$ provided f is sufficiently regular. We will discuss this matter later on.

2. In the cases $n = 3,4$ a weak solution satisfying all our assumptions is furnished by the Galerkin approximation procedure. This was stated in theorem IV.1.2.

3. The first part of the present theorem was proved in [SW] under the slightly different assumption $f \in L^2((0,+\infty),(L^n(\Omega))^n)$ and $u \in L^q((0,+\infty),(L^n(\Omega))^n)$ for some $q > 1$. These assumptions were made in order to show that u is regular for large times; it is easily seen from the proof in [SOW1] that the decomposition (IV.5.10) holds under the assumptions of the present theorem.

Proof of theorem IV.5.5: Let

$$N = \{s \mid s \in (0,+\infty),\ u(s) \in (L^n(\Omega))^n,$$
$$\text{(IV.5.9) holds in } s \text{ and for all } t \geq s\}.$$

Of course $(0,+\infty)-N$ has measure 0. For any $s \in N$, according to theorem III.3.6, we can construct the local strong solution $\tilde{u}$ of the Navier-Stokes equations on $[s,s+T(u(s)))$ with $\tilde{u}(s) = u(s)$ and inhomogeneous term $P_p f$. Since (IV.5.9) holds in s for all

$t \geq s$ we get with the aid of theorems IV.4.1, IV.3.1 that $u(t) = \tilde{u}(t)$, $s \leq t < s+T(u(s))$. Now we consider the set

$$(IV.5.13) \quad \bigcup_{s \in N} [s, s+T(u(s))) \supset N.$$

As in the proof of theorem IV.5.4 we see that the components of connectedness are at most countable since they consist of intervals of positive length. Let us enumerate them by $I_1, I_2, \ldots$. Then for any I_λ we get that $u(s) \in (L^n(\Omega))^n$, $s \in I_\lambda$, and (IV.5.9) holds in s, $s \in I_\lambda$, and for all $t \geq s$. On $\overset{\circ}{I}_\lambda$ then the procedure of the proof of theorem IV.5.4 is applied (Observe that to do so we only need the energy inequality everywhere and $u(s) \in (L^n(\Omega))^n$ in $\overset{\circ}{I}_\lambda$). This gives

$$\overset{\circ}{I}_\lambda = \bigcup_{\nu=1}^{\infty} J_{\nu,\lambda} \cup S_\lambda$$

where the $J_{\nu,\lambda}$ are open intervals with

$$u \in C^o(J_{\nu,\lambda}, (H^{2,p}(\Omega))^n),$$

$$J_{\nu,\lambda} \cap J_{\mu,\lambda'} = \emptyset, \quad (\nu,\lambda) \neq (\mu,\lambda'),$$

and where S_λ is countable. Setting a_λ, b_λ for the endpoints of I_λ we see that

$$(0,+\infty) = \bigcup_{\lambda=1}^{\infty} I_\lambda \cup ((0,+\infty) - \bigcup_{\lambda=1}^{\infty} I_\lambda)$$

$$= \bigcup_{\lambda=1}^{\infty} \overset{\circ}{I}_\lambda \cup [((0,+\infty) - \bigcup_{\lambda=1}^{\infty} I_\lambda) \cup \{a_\lambda, b_\lambda\}]$$

$$= \bigcup_{\lambda=1}^{\infty} \bigcup_{\nu=1}^{\infty} J_{\nu,\lambda} \cup [((0,+\infty) - \bigcup_{\lambda=1}^{\infty} I_\lambda) \cup \{a_\lambda, b_\lambda\} \cup \bigcup_{\lambda=1}^{\infty} S_\lambda]$$

if all b_λ are $< +\infty$. Enumerating the $J_{\nu,\lambda}$ we get the J_ν. Since (IV.5.13) holds the set

$$S = [((0,+\infty) - \bigcup_{\lambda=1}^{\infty} I_\lambda) \cup \{a_\lambda, b_\lambda\} \cup \bigcup_{\lambda=1}^{\infty} S_\lambda]$$

has measure 0 and moreover

$$S = (0,+\infty) - \bigcup_{\nu=1}^{\infty} J_\nu .$$

Thus the first part of part 1. of our theorem is proved. For the second part of 1. we choose a $T_1 \geq \tilde{T}_1$ with:

$$u(T_1) \in (L^n(\Omega))^n,$$

the energy inequality (IV.5.9) holds in T_1 and for all $t \geq T_1$. Then we consider the local strong solution $\tilde{u}$ on $[T_1, T_1+T_1(u(t_1)))$ with initial value $u(T_1)$ and inhomogeneous term $P_n f$ in the sense of theorem III.3.6. If

$$\|P_n f(t)\|_{H_n(\Omega)} + \|u(T_1)\|_{H_n(\Omega)} \leq \varepsilon_1,$$

where ε_1 is a certain positive quantity depending on A_n then according to theorem III.4.1, part 4., we have $T(u(t_1)) = +\infty$. Theorem IV.4.1 completes the proof of the first part of the present theorem. As for the second part we get first from the energy inequality

$$\|u(t)\|^2_{(L^2(\Omega))^n} + 2\nu \int_s^t \|\nabla u(\sigma)\|^2_{(L^2(\Omega))^{n^2}} d\sigma$$

$$\leq \|u(s)\|^2_{(L^2(\Omega))^n} + \varepsilon \int_s^t \|\nabla u(\sigma)\|^2_{(L^2(\Omega))^{n^2}} d\sigma +$$

$$+ c(\varepsilon) \int_s^t \|f(\sigma)\|^2_{(H^{-1,2}(\Omega))^n} d\sigma$$

for some $s > 0$, for all $t \geq s$ and for all $\varepsilon > 0$. This proves that

$$\text{(IV.5.14)} \quad u \in L^2((0,+\infty), (\overset{\circ}{H}{}^{1,2}(\Omega))^n).$$

Since by Sobolev

(IV.5.15) $(L^6(\Omega))^3 \supset (\overset{\circ}{H}{}^{1,2}(\Omega))^3$, $n = 3$,

(IV.5.16) $(L^4(\Omega))^4 \supset (\overset{\circ}{H}{}^{1,2}(\Omega))^4$, $n = 4$

with continuous imbeddings we see that in any case

$$(L^n(\Omega))^n \supset (\overset{\circ}{H}{}^{1,2}(\Omega))^n$$

with a continuous imbedding. Because of (IV.5.14) we get that $\|u(T_1)\|_{H_n(\Omega)}$ can be made arbitrarily small if T_1 is sufficiently large. Together with (IV.5.12) it follows from part 1. that the second part of 2. of our theorem is proved. As for the first part of 2. it readily follows from 1. if one takes into consideration (IV.5.15), (IV.5.16). Our theorem is proved. □

As it is seen from the proof of part 2. of theorem IV.5.5 in the space dimension 3 already theorem III.3.4 is sufficient to prove Leray's structure theorem: Theorem III.3.4 works with initial values from $H_p(\Omega)$ for some $p > n$ however, but in view of (IV.5.15) we can choose $p = 6$. The following theorem shows that Leray's structure theorem is valid under assumptions on f being slightly different from those we have made in theorem IV.5.5; moreover it characterises the validity of Leray's structure theorem in terms of the regularity of u where u is any weak solution under consideration. The proof, however, has a conceptual advantage in as much as it makes no use at all of the construction of local strong solutions with bad initial values.

Theorem IV.5.6: Let $n = 3$. 1. Let $\varphi \in H_2(\Omega)$, let $f \in C^{0,1}([0,T], (L^{5/4}(\Omega))^3) \cap C^{\frac{1}{2p}}([0,T], (L^p(\Omega))^n)$ for some $p > n = 3$ and for all $T > 0$. Let

$$f \in L^{5/4}((0,+\infty), (L^{5/4}(\Omega))^n).$$

Let

$$\|P_{\frac{5}{4}}f(t)\|_{H_{5/4}(\Omega)} + \|P_{\frac{5}{4}}f'(t)\|_{H_{5/4}(\Omega)} \leq \varepsilon_1$$

for all $t \geq \tilde{T}_1$ for some $\tilde{T}_1 \geq 0$, where ε_1 is a certain positive number depending on $A_{5/4}$. Let u be a weak solution of the Navier-Stokes equations over $(0,T)\times\Omega$ for all $T>0$ in the sense of definition IV.1.1 with boundary values $u(0)=\varphi$, $u|\partial\Omega = 0$ and inhomogeneous term f. Moreover u is supposed to fulfill the energy inequality almost everywhere, i.e.

$$\|u(t)\|^2_{H_2(\Omega)} + 2\nu \int_s^t \|\nabla u(\sigma)\|^2_{(L^2(\Omega))^{n^2}} \, d\sigma$$

$$\leq \|u(s)\|^2_{H_2(\Omega)} + 2 \int_s^t (f(\sigma),u(\sigma)) \, d\sigma$$

for almost all $s>0$ and all $t \geq s$. Then Leray's structure theorem holds for u, i.e.:

$$(0,+\infty) = \bigcup_{\nu=1}^{\infty} J_\nu \cup S,$$

where the open intervals J_ν are pairwise disjoint, where $u \in C^0(J_\nu,(C^{1+\alpha}(\bar{\Omega}))^n)$ for some $\alpha \in (0,1)$, where

$$S \cup \{0\} = [0,+\infty) - \bigcup_{\nu=1}^{\infty} J_\nu$$

is a compact set of measure 0 and where one of the J_ν is semi-infinite, i.e.

$$u \in C^0([T_1,+\infty),(C^{1+\alpha}(\bar{\Omega}))^n)$$

for some $T_1 > 0$.

2. Let φ, f be as in part 1. Let u be an arbitrary weak solution over $(0,T)\times\Omega$ for all $T>0$ with boundary values $u(0)=\varphi$, $u|\partial\Omega=0$ and inhomogeneous term f. Then Leray's structure theorem holds

<u>for</u> u <u>if and only if</u>:

$$u\cdot u \in L^2((t,t+\varepsilon(t)),(L^2(\Omega))^{n^2})$$

<u>for almost all</u> $t>0$ <u>and some</u> $\varepsilon(t)>0$ <u>which may depend on</u> t, <u>and moreover</u>

$$u\cdot u \in L^2((\tilde{T}_1,T),(L^2(\Omega))^{n^2})$$

<u>for the</u> $\tilde{T}_1>0$ <u>from</u> 1. <u>and all</u> $T>\tilde{T}_1$.

<u>Proof:</u> First we observe that $\frac{5}{4}>\frac{n}{3}=1$. From theorem IV.2.2 and the remark after the proof of theorem IV.3.3 it follows that

$$u(t)\in (H^{2,5/4}(\Omega))^n \cap (\overset{\circ}{H}{}^{1,5/4}(\Omega))^n \cap H_{5/4}(\Omega) = D(A_{5/4})$$

a.e. on $(0,+\infty)$. Thus for almost all t we can construct a strong solution $\tilde{u}$ of the Navier-Stokes equations

$$\text{(IV.5.16)}\quad \tilde{u}' + A_{\frac{5}{4}}u + M(\tilde{u}) = P_{\frac{5}{4}}f,$$

$$\tilde{u}(t) = u(t)$$

in the sense of theorem III.3.1 on the interval $[t,t+T(u(t)))$. As it was proved in theorem III.3.2 the strong solution $\tilde{u}$ is in $C^{1+\alpha}(\bar{\Omega})$ for some $\alpha\in(0,1)$ and

$$\text{(IV.5.17)}\quad \|\nabla\tilde{u}(s)\|_{C^{\alpha}(\bar{\Omega})} \le \frac{c(\tilde{T})}{(s-t)^{1-\rho}},\quad t<s\le\tilde{T}<T(u(t))$$

for some $\rho\in(0,1)$. Thus taking the scalar product of (IV.5.16) with $\tilde{u}$ over $[t+\varepsilon,s]$, $t<t+\varepsilon<s<T(u(t))$, we arrive at ($\langle .,.\rangle$ = $L^{5/4}$-L^5 duality)

$$\int_{t+\varepsilon}^{s} \langle\tilde{u}'(\sigma),\tilde{u}(\sigma)\rangle\, d\sigma + \nu\int_{t+\varepsilon}^{s} \|\nabla\tilde{u}(\sigma)\|_{(L^2(\Omega))^{n^2}}\, d\sigma =$$

$$= \int_{t+\varepsilon}^{s} \langle P_{\frac{5}{4}} f(\sigma), \tilde{u}(\sigma)\rangle \, d\sigma.$$

Since $\tilde{u} \in C^1([t,t+T(u(t))),(L^{5/4}(\Omega))^n) \cap C^0([t,t+T(u(t))),$ $(H^{2,5/4}(\Omega))^n \cap (\overset{\circ}{H}{}^{1,5/4}(\Omega))^n \cap H_{5/4}(\Omega))$ it may be shown by approximation of u in the space $C^1([t+\varepsilon,s],(L^{5/4}(\Omega))^n)$ that

$$\int_{t+\varepsilon}^{s} \langle \tilde{u}'(\sigma), \tilde{u}(\sigma)\rangle \, d\sigma = \frac{1}{2}\|\tilde{u}(s)\|^2_{(L^2(\Omega))^n} - \frac{1}{2}\|\tilde{u}(t+\varepsilon)\|^2_{(L^2(\Omega))^n}$$

and consequently

$$\text{(IV.5.18)} \quad \|\tilde{u}(s)\|^2_{(L^2(\Omega))^n} + 2\nu \int_{t+\varepsilon}^{s} \|\nabla\tilde{u}(\sigma)\|^2_{(L^2(\Omega))^{n^2}} \, d\sigma =$$

$$= \|\tilde{u}(t+\varepsilon)\|^2_{(L^2(\Omega))^n} + 2 \int_{t+\varepsilon}^{s} (f(\sigma), u(\sigma)) \, d\sigma.$$

As it has already been exploited for the proof of theorem III.3.2 we have by Sobolev $(H^{2,5/4}(\Omega))^n \subset (L^{15/2}(\Omega))^n$ with a continuous imbedding and therefore $\tilde{u} \in C^0([t,t+T(u(t)),(L^{15/2}(\Omega))^n)$. This furnishes (IV.5.18) also for $\varepsilon = 0$, i.e. the energy equality holds for $\tilde{u}$ in t and all s, $t \leq s \leq T(u(t))$. It may be left the reader to show that $\tilde{u}$ is a weak solution of the Navier-Stokes equations over $(t,t+T(u(t)))\times\Omega$ with boundary values $\tilde{u}(t) = u(t)$, $u|\partial\Omega = 0$ in the sense of definition IV.1.1. Theorem IV.4.1 shows that

$$u(\sigma) = \tilde{u}(\sigma), \quad t \leq \sigma < t+T(u(t)).$$

For the weak solution u under consideration we get for $\tilde{T}_1 < t$ (cf. theorem IV.2.2, we assume $u(\tilde{T}_1) \in (H^{2,5/4}(\Omega))^n \cap$ $(\overset{\circ}{H}{}^{1,5/4}(\Omega))^n \cap H_{5/4}(\Omega)$)

$$\text{(IV.5.19)} \quad \int_{\tilde{T}_1}^{t} \|u'(\sigma)\|^{5/4}_{(L^{5/4}(\Omega))^n} \, d\sigma + \int_{\tilde{T}_1}^{t} \|u(\sigma)\|^{5/4}_{(H^{2,5/4}(\Omega))^n} \, d\sigma$$

$$+ \sup_{\tilde{T}_1 \leq \sigma \leq t} \|u(\sigma)\|^{5/4}_{(W^{2/5,5/4}(\Omega))^n} \leq$$

$$\leq \hat{c}\Big(\int_{\tilde{T}_1}^{t} \|f(\sigma)\|^{5/4}_{(L^{5/4}(\Omega))^n}\, d\sigma + \int_{\tilde{T}_1}^{t} \|u\cdot\nabla u\|^{5/4}_{(L^{5/4}(\Omega))^n}\, d\sigma$$

$$+ \|u(\tilde{T}_1)\|^{5/4}_{(W^{2/5,5/4}(\Omega))^n}\Big)$$

if we apply Solonnikov's estimate in a form being slightly different from (III.1.4) on $(\tilde{T}_1,t)$ instead of $(0,T)$; the main point here is that the constant $\hat{c}$ on the right side does not depend on t and the term $\int_{\tilde{T}_1}^{t} \|u(\sigma)\|^{5/4}_{(L^{5/4}(\Omega))^n}\, d\sigma$ is missing. The justification for the estimate (IV.5.19) is given as follows: First of all note that from the energy inequality for u it follows as in the proof of theorem IV.5.5 (see (IV.5.14)) that

$$u \in L^2((0,+\infty),(\overset{o}{H}{}^{1,2}(\Omega))^n) \cap L^\infty((0,+\infty),(L^2(\Omega))^n).$$

According to the estimates in the proof of proposition IV.2.1 we obtain that

$$u\cdot\nabla u \in L^{5/4}((0,+\infty),(L^{5/4}(\Omega))^n).$$

Setting $F(t) = f(t)-u\cdot\nabla u$ we have $u' - \nu\Delta u + \nabla\pi = F(t)$, $\nabla\cdot u = 0$, $u(0) = \varphi$, and consequently

$$u'(t) + A_{5/4}u(t) = P_{5/4}F(t) \text{ a.e. on } (\tilde{T}_1,+\infty),$$

$$\text{initial value} = u(\tilde{T}_1) \in D(A_p)$$

(cf. theorem IV.2.2). Approximating F in $L^{5/4}((\tilde{T}_1,t),(L^{5/4}(\Omega))^n)$ by $F_k \in C^1([\tilde{T}_1,t],(L^{5/4}(\Omega))^n)$ we see that the solutions (w_k,π_k)

$$w_k' - \nu\Delta w_k + \nabla\pi_k = F_k,\ \nabla\cdot w_k = 0,\ w_k(\tilde{T}_1) = u(\tilde{T}_1),$$

according to theorem III.1.1, converge in the norms $|||(.,.)|||_{5/4,(\tilde{T}_1,t)}$ and in particular in $C^0([\tilde{T}_1,t],(L^{5/4}(\Omega))^n)$. Since $w_k'+A_{5/4}w = PF_k$, $w_k(\tilde{T}_1) = u(\tilde{T}_1)$, by this limit process we arrive at

$$u(t) = e^{-tA_{5/4}}u(\tilde{T}_1) + \int_{\tilde{T}_1}^{t} e^{-(t-s)A_{5/4}}F(s)\,ds,\ t \geq \tilde{T}_1.$$

Using $\|e^{-(t-s)A_{5/4}}x\|_{H_{5/4}(\Omega)} \leq ce^{-\delta(t-s)}\|x\|_{H_{5/4}(\Omega)}$ (cf. (III.1.15)) we get with a constant being independent from t

$$\int_{\tilde{T}_1}^{t} \|u(\sigma)\|_{H_{5/4}(\Omega)}^{\frac{5}{4}}\,d\sigma \leq c(\|u(\tilde{T}_1)\|_{H_{5/4}(\Omega)} + \int_{\tilde{T}_1}^{t} \|F(\sigma)\|_{H_{5/4}(\Omega)}^{5/4}\,d\sigma).$$

Thus (IV.5.19) is correct. Letting t tend to $+\infty$ we obtain in particular that

$$\int_{\tilde{T}_1}^{\infty} \|u(\sigma)\|_{H^{2,\frac{5}{4}}(\Omega)}^{\frac{5}{4}}\,d\sigma < +\infty.$$

Our assumption $u(\tilde{T}_1) \in D(A_{5/4})$ is in fact no restriction since we already know that $u(t) \in D(A_{5/4})$ a.e. (cf. theorem IV.2.2). Thus for a $T_1 > \tilde{T}_1$ the quantity $\|A_{5/4}u(T_1)\|$ can be made arbitrarily small. Constructing again the local strong solution $\tilde{u}$ on $[T_1, +\infty)$ with $\tilde{u}(T_1) = u(T_1)$ we see that according to theorem III.4.1, 1. we get $T(u(T_1)) = +\infty$. We know already from the first part of the proof that $u(t) = \tilde{u}(t)$, $t \geq T_1$. This gives the second part of the present theorem, 1..As for the first part of 1. we can proceed as in the proof of theorem IV.5.5. Now let us turn to part 2. of our theorem. If the structure theorem holds then clearly

$$\text{(IV.5.20)}\quad u\cdot u \in L^2((t,t+\varepsilon(t)),(L^2(\Omega))^{n^2})$$

for almost all $t > 0$ and some $\varepsilon(t) > 0$. On the other hand, if (IV.5.20) holds, then we can construct the strong solution $\tilde{u}$ on $[t, t+T(u(t)))$ with $\tilde{u}(t) = u(t)$ for all t being contained in the set

$$N = \{t \mid u(t) \in D(A_{\frac{5}{4}}),\ u\cdot u \in L^2((T,t+\varepsilon(t)),(L^2(\Omega))^{n^2}) \text{ for some } \varepsilon(t) > 0$$

then

$$\tilde{u}(t) = u(t) \text{ on } [t,t+\varepsilon(t)) \cap [t,t+T(u(t))).$$

This follows from theorems IV.3.3, IV.4.1 or by the more direct approach in theorem IV.4.3; clearly $[0,+\infty)-N$ has measure 0. For a $T_1 \geq \tilde{T}_1$ with

$$u(T_1) \in D(A_p)$$

we can argue as follows: We have

$$u\cdot u \in L^2((T_1,T),(L^2(\Omega))^{n^2})$$

for all $T > T_1$. Therefore we get for the local strong solution $\tilde{u}$ on $[T_1,T_1+T(u(T_1)))$ with initial value $\tilde{u}(T_1) = u(T_1)$ that $\tilde{u}(t) = u(t)$ on $[T_1,T_1+T_1(u(T_1))$. In the same way as in the first part it is shown that $T(u(T_1)) = +\infty$ if T_1 is sufficiently large. The rest of the proof can be completed in the same way we have proved theorem IV.5.5. □

Remark: 1. The preceding theorem shows that approximately Leray's structure theorem is equivalent with the square integrability of $u\cdot u$ at almost every $t > 0$ and at $t = +\infty$ $(n = 3)$.

2. S is denoted the singular set of u and the weak solutions we have constructed in the two preceding theorems frequently are called turbulent solutions.

3. As for the inhomogeneous term it is sufficient to assume that $f \in L^\infty((0,T),(L^n(\Omega))^n)$ for all $T > 0$. This will be clear from a careful inspection of the proof of theorem III.3.2. We have made the assumption $f \in C^{1/2p}([0,T],(L^p(\Omega))^n)$ only in order to have the formal connection to theorem III.3.2. □

Finally we deal with case $n=2$. In this case it turns out that any weak solution over $(0,+\infty)\times\Omega$ is a regular solution for $t>0$.

Theorem IV.5.7: Let $n=2$. Let $\varphi\in H_2(\Omega)$, let $f\in C^{\frac{1}{2p}}([0,T],(L^p(\Omega))^n)\cap C^0((0,T],(H^{\frac{1}{p},p}(\Omega))^n)$ for some $p>n=2$ and all $T>0$. Let u be a weak solution of the Navier-Stokes equations in the sense of definition IV.1.1 over $(0,T)\times\Omega$ for all $T>0$ with boundary values $u(0)=\varphi$, $u|\partial\Omega=0$ and inhomogeneous term f. Then

$$u\in C^0((0,+\infty),(H^{2,p}(\Omega))^n).$$

Proof: Since $u\in L^\infty((0,T),(L^2(\Omega))^n)$, $T>0$, the weak solution fulfills the energy equality everywhere and thus we have

$$u\in C^0([0,T],(L^2(\Omega))^2) = \\ = C^0([0,T],(L^n(\Omega))^n)$$

(theorem IV.3.1). Also u is uniquely determined by its data φ,f (theorem IV.4.1). Thus constructing the strong solution $\tilde{u}$ on $[0,T(\varphi))$ according to theorem III.3.6 we get

$$\text{(IV.5.21)}\quad u(t)=\tilde{u}(t) \text{ on } [0,T(\varphi))$$

since $\tilde{u}$ is also a weak solution over $(0,T(\varphi))\times\Omega$ (see the proof of theorem IV.4.2). Let us assume that $T(\varphi)<+\infty$. (IV.5.21) shows that $\tilde{u}$ is uniformly continuous on $[0,T(\varphi))$ with respect to t in the $H_2(\Omega)$-norm; according to theorem III.3.6 this is a contradiction to our assumption $T(\varphi)<+\infty$. Thus $T(\varphi)=+\infty$ and our theorem is proved. □

Remark: It may be noted that in the case of dimension 2 there is a global existence and regularity result for $p\geq 2$ being valid under the assumptions on φ,f as they have been made in theorem III.3.1; also it is then not necessary to reconstruct the weak solution as a local strong one. This was proved in [W10]; in this proof however the assumption $\varphi \in (H^{2,p}(\Omega))^n \cap (\overset{\circ}{H}{}^{1,p}(\Omega))^n \cap H_p(\Omega)$ was made but it can readily be seen from the proof that $\varphi \in (\tilde{W}^{2-2/p,p}(\Omega))^n$, $\nabla\cdot\varphi = 0$ is sufficient.

§ 6. Comments to chapter IV

In chapter IV we have already made many comments in the remarks. Thus we will be brief in most cases.

To § 1: It is clear that all our considerations here are valid for any dimension $n \geq 2$. We have to concentrate on real φ,f and real weak solutions u in the end of § 1 since otherwise the relations $(u\cdot\nabla u,u) = 0$ or $(v\cdot\nabla u,u) = 0$ could not be used.

To § 2: All our considerations are again valid for arbitrary n if one takes into consideration the n-dimensional analogue of theorem III.1.1. φ,f and u may be complex valued also. The basic idea of this paragraph goes back to [L].

To § 3: The idea to connect the energy (in)equality with the question of uniqueness and regularity of a weak solution under consideration is expounded in [Ser2]. The study of the regularity of the pressure π was motivated by a recent paper of Caffarelli, Kohn and Nirenberg [CKN]; as it is shown in this paper the properties of π are initimately connected with partial regularity of a weak solution u considered as a function of t,x $(n = 3)$. The results in [CKN] are most sharply in the case of the Cauchy-problem $(\Omega = \mathbb{R}^3)$. The result indicated in § 3 in this case follows quite simply from the foregoing considerations by continuing u' to $(H^{1,\frac{n}{2}}(\Omega))^n$ by O; u' then is still the distributional derivative of u with respect to t, and the notion of a weak solution for $\Omega = \mathbb{R}^n$ is defined analogously to IV.1.1. Having some additional information on π one can also show

$$\pi \in L^{\frac{n+2}{n}}((0,T),L^{\frac{n+2}{n}}(\mathbb{R}^n))$$

(cf. [CKN]). In fact this is, to some extent, also true for exterior domains Ω instead of $\mathbb{R}^n$. This is proved in a paper by H. Sohr and the author, [SOW2]; there conclusions analogous to [CKN] will be drawn from this fact in order to consider the partial regularity of a weak solution. All our considerations in § 3 are valid for arbitrary n.

<u>To § 4:</u> The basic idea of the proof is due to Serrin [Ser2], namely to employ the energy inequality. We generalise Serrin's theorem in two respects, namely as it concerns the dimension n and then as it concerns the admissible function classes for u^1. With the exception of theorem IV.4.5 all our considerations are valid for arbitrary n (cf. the comments to III.).

<u>To § 5:</u> The results from theorem IV.5.4 on deal with the singular points in t. The idea to do so goes back to [Ler]. Some part of the material spread out here was presented in [SOW1]. In this respect we want to mention recent preprints by Giga [Gi3] and Masuda [M]. Giga has proved a theorem similar to IV.5.5 by a method being somewhat different from ours; Masuda has also proved the uniqueness theorem being basic for our considerations here, and he moreover treats unbounded domains.

V. Global Solutions of Abstract Nonlinear Parabolic Equations and Applications

This chapter deals first with the question under which conditions a local strong solution of an abstract equation $u' + Au + M(u) = f$, $u(0) = \varphi$, in a reflexive Banach space B can be continued to the whole positive real axis; here $-A$ generates an analytic semigroup e^{-tA}. Secondly we want to apply our theory to nonlinear parabolic systems and to the Navier-Stokes equations.

§ 1. Abstract Nonlinear Parabolic Equations

Let B be a reflexive Banach space, let A be a closed operator whose resolvent set is contained in $\{\lambda \mid \lambda \in \mathbb{C},\ \mathrm{Re}\ \lambda \geq 0\}$. Moreover we assume that D(A) is dense and that

$$||(\lambda+A)^{-1}|| \leq \frac{\hat{M}}{|\lambda|+1},\ \mathrm{Re}\ \lambda \geq 0,$$

for some positive constant $\hat{M}$. Thus $-A$ generates an analytic semigroup e^{-tA} with

$$||e^{-tA}|| \leq \hat{M}e^{-\delta t}$$

for some $\delta > 0$; also the fractional powers of A may be defined (see I. for all this material).

We assume that M is a mapping from $[0,T]\times D(A)$ into B fulfilling essentially the Lipschitz condition of II.1, namely

(V.1.1) $\|M(t,u)-M(s,v)\| \leq k(T,\|Au\|+\|Av\|)(|t-s|+\|A^{1-\rho}(u-v)\|)$,

$$0 \leq t,s \leq T, \quad u,v \in D(A)$$

for some $\rho \in (0,1)$ and some continuous function $k:[0,+\infty)\times[0,+\infty) \to [0,+\infty)$. According to theorem II.1.1 we then can construct a local strong solution of

$$u' + Au + M(t,u) = 0,$$
$$u(0) = \varphi \in D(A)$$

on a maximal interval $[0,T(\varphi)) \subset [0,T]$ which has the following properties: $u \in C^1([0,T(\varphi)),B)$, $u(t) \in D(A)$, $Au \in C^0([0,T(\varphi)),B)$, $u'(t) \in D(A^{1-\rho})$ for ρ as above, $0 < t < T(\varphi)$, $A^{1-\rho}u' \in C^0((0,T(\varphi)),B)$, $\|A^{1-\rho}u'(t)\| \leq \frac{c(\tilde{T})}{t^{1-\rho}}$, $0 < t \leq \tilde{T} < T(\varphi)$. If $T(\varphi) < T$ then

$$\lim_{t \uparrow T(\varphi)} \|Au(t)\| = +\infty.$$

We want to give a sufficient criterion in order to guarantee that our local solution can be continued on $[0,T]$.

I. _Assumptions on the nonlinearity:_ _There is a Banach space_ V _such that_

(V.1.2) $D(A) \subset V$

with a continuous imbedding (D(A) _is normed by_ $\|A.\|$) _and_

(V.1.3) $\|M(t,u)\| \leq \hat{k}(T,\|u\|_V)\|Au\|$,

$0 \leq t \leq T$, $u \in D(A)$, _with a continuous function_ $\hat{k}$ _as_ k _above_. _Moreover_ M(t,u) _is supposed to fulfill the following Lipschitz condition:_

(V.1.4) $\|M(t,u)-M(s,v)\| \leq \|A(u-v)\|g_1(\|u-v\|_V+|t-s|)\hat{k}(T,\|u\|_V+\|v\|_V) +$

$$+ (\|A(u-v)\|^{1-\varepsilon}+1)g_2(s,\|Av\|+\|u\|_V+\|v\|_V).$$

The "remainder" $(\|A(u-v)\|^{1-\varepsilon}+1)g_2(s,\|Av\|+\|u\|_V+\|v\|_V)$ is denoted by $R(s,u,v)$. ε is some number from $(0,1)$, $g_1:[0,+\infty)\to[0,+\infty)$, $g_2:[0,+\infty)\times[0,+\infty)\to[0,+\infty)$ are continuous functions with $g_1(x)\to 0$, $x\to 0$. Finally we assume that for $t\in[0,\tilde T]$, $0<\tilde T<T$,

$$(V.1.5)\quad \|M'(t,\tilde u(t))\| \le c(\|A\tilde u(t)\|+1)^N \|A^{1-\rho}\tilde u'(t)\|$$

where $\rho\in(0,1)$ is the number from (V.1.1) and where $\tilde u$ is any function with $\tilde u(t)\in D(A)$, $0\le t\le T$, $A\tilde u\in C^0([0,\tilde T],B)$, $\tilde u\in C^1((0,\tilde T],B)$, $\tilde u'(t)\in D(A^{1-\rho})$, $A^{1-\rho}u'\in C^0((0,\tilde T],B)$, $\|A^{1-\rho}u'(t)\|\le\frac{c(\tilde T)}{t^{1-\rho}}$, $0<t\le\tilde T<T$. N is some element from $\mathbb{N}$ and M' is the derivative of M in the weak sense (cf. 0.2, p. VII, Lemma II.1.1).

II. Assumptions on the linear part: For any solution u of

$$u' + Au = f,\quad u(0)=\varphi\in D(A)$$

on $[0,\tilde T]$ for some $\tilde T$, $0<\tilde T\le T$, we have the estimate

$$(V.1.6)\quad \int_0^{\tilde T}\|u'(t)\|^q\,dt + \int_0^{\tilde T}\|Au(t)\|^q\,dt \le c\Big(\int_0^{\tilde T}\|f(t)\|^q\,dt + \|A\varphi\|^q\Big),\qquad q>\max(2,\tfrac{N}{\rho}),$$

where the constant c may depend on q but not on $\tilde T$. f is assumed to be in $C^0([0,\tilde T],B)$, u to be in $C^0([0,T],D(A))\cap C^1([0,T],B)$.

Remark: (V.1.5) is superfluous if $\|M(t,u)-M(s,v)\|\le k(T,\|A^{1-\rho}u\|+\|A^{1-\rho}v\|)(|t-s|+\|A^{1-\rho}(u-v)\|)$ instead of (V.1.1). The estimate (V.1.6) is then only needed for $q\ge q_0=q_0(\rho)$ with some $q_0(\rho)>1$ depending on ρ alone.

Our theorem on global solutions then is.

Theorem V.1.1: *Let* (V.1.1) *and assumptions* I., II. *be fulfilled. Let* u *be the local strong solution of*

$$(V.1.7)\quad u' + Au + M(t,u) = 0, \qquad u(0) = \varphi$$

already having been constructed in theorem II.1.1. *If* $T(\varphi) < T$, *then*

$$u:[0,T(\varphi)) \to V$$

is not uniformly continuous. If $T(\varphi) = T$ *and*

$$u:[0,T) \to V$$

is uniformly continuous then u *may be continued on* [0,T], *i.e.* $u \in C^1([0,T],B)$, $Au \in C^0([0,T],B)$, *and* u *fulfills* (V.1.7) *on* [0,T].

Proof: We assume that $T(\varphi) < T$ and that

$$u:[0,T(\varphi)) \to V$$

is uniformly continuous. We divide $[0,T(\varphi)]$ into possibly small intervals of constant length:

$$[0,T(\varphi)] = \bigcup_{j=1}^{\hat{N}} [t_j, t_{j+1}],$$

$t_1 = 0$, $t_{\hat{N}+1} = T(\varphi)$. $|t_{j+1} - t_j| = \delta$ will be made small later on. For $\tilde{T} = \delta \cdot \tilde{N} = t_{\tilde{N}+1}$ we get

$$\int_0^{\tilde{T}} \|u'(t)\|^q \, dt + \int_0^{\tilde{T}} \|Au(t)\|^q \, dt =$$

$$= \sum_{j=1}^{\tilde{N}} \left(\int_{t_j}^{t_{j+1}} \|u'(t)\|^q \, dt + \int_{t_j}^{t_{j+1}} \|Au(t)\|^q \, dt \right),$$

$$\leq c(q) \sum_{j=1}^{\tilde{N}} \left(\int_{t_j}^{t_{j+1}} \|M(t,u(t))\|^q \, dt + \|A\varphi|^q\right),$$

$$\leq c(q) \sum_{j=1}^{\tilde{N}} \int_{t_j}^{t_{j+1}} \|M(t,u(t)) - M(t_j,u(t_j))\|^q \, dt +$$

$$+ c(q)\left(\sum_{j=1}^{\tilde{N}} \int_{t_j}^{t_{j+1}} \|M(t_j,u(t_j))\|^q \, dt + \|A\varphi|^q\right),$$

$$\leq c(q)\Big\{\sum_{j=1}^{\tilde{N}} \Big(\int_{t_j}^{t_{j+1}} \|A(u(t)-u(t_j))\|^q g_1^q(\|u(t)-u(t_j)\|_V + |t-t_j|)\cdot$$

$$\cdot \hat{k}^q(T,\|u(t)\|_V + \|u(t_j)\|_V)\,dt +$$

$$+ \int_{t_j}^{t_{j+1}} (\|A(u(t)-u(t_j))\|^{1-\varepsilon} + 1)^q \cdot$$

$$\cdot g_2^q(t_j,\|Au(t_j)\| + \|u(t)\|_V + \|u(t_j)\|_V)\,dt +$$

$$+ \int_{t_j}^{t_{j+1}} \|M(t_j,u(t_j))\|^q \, dt\Big) + \|A\varphi|^q\Big\},$$

$$\leq c(q)\Big\{\sum_{j=1}^{\tilde{N}} \Big(\int_{t_j}^{t_{j+1}} \|Au(t)\|^q g_1^q(\|u(t)-u(t_j)\|_V + |t-t_j|)\cdot$$

$$\cdot \hat{k}^q(T,\|u(t)\|_V + \|u(t_j)\|_V)\,dt +$$

$$+ \int_{t_j}^{t_{j+1}} \|Au(t_j)\|^q g_1^q(\|u(t)-u(t_j)\|_V + |t-t_j|)\cdot$$

$$\cdot \hat{k}^q(T,\|u(t)\|_V + \|u(t_j)\|_V)\,dt +$$

$$+ \varepsilon' \int_{t_j}^{t_{j+1}} (\|Au(t)\|+1)^q g_2^q(t_j,\|Au(t_j)\| + \|u(t)\|_V + \|u(t_j)\|_V)\,dt +$$

$$+ c(\varepsilon') \int_{t_j}^{t_{j+1}} (\|Au(t_j)\|+1)^q g_2^q(t_j,\|Au(t_j)\| +$$

$$+ \|u(t)\|_V + \|u(t_j)\|_V)\,dt +$$

$$+ \int_{t_j}^{t_{j+1}} k^q(T,\|u(t_j)\|_V)\|Au(t_j)\|^q \, dt\Big) + \|A\varphi|^q\Big\},\ 0 < \varepsilon'.$$

Since $u:[0,T(\varphi))\to V$ is uniformly continuous u can be continued in $T(\varphi)$ in a unique way as a continuous function. Thus

$$\sup_{0\le t<T(\varphi)} \|u(t)\|_V \le D.$$

Choosing δ that small that with $\hat{k}_D = \sup_{0\le x\le 2D} \hat{k}(T,x)$

$$\text{(V.1.8)}\qquad \sup_{\substack{|t-t_j|\le\delta,\\ 1\le j\le N,\\ 0\le t<T(\varphi)}} c(q)\cdot g_1^q(\|u(t)-u(t_j)\|_V+|t-t_j|) \le \frac{1}{4k_D}$$

and choosing $\varepsilon' = \frac{1}{4c(q)}$ we see that

$$\text{(V.1.9)}\qquad \int_{t_{\tilde N}}^{t_{\tilde N+1}} \|u'(t)\|^q\,dt + \int_{t_{\tilde N}}^{t_{\tilde N+1}} \|Au(t)\|^q\,dt,\ q\ge 2$$

is estimated a priori in dependence of q and

$$D,\ \sup_{0\le t\le t_{\tilde N}} \|Au(t)\|.$$

We want to emphasize at this stage of the proof that the smallness of δ needed in (V.1.8) does not depend on $t_{\tilde N}$. Now we use the estimates for our local strong solution coming from the integral equation. We replace φ by $u(t_{\tilde N})$ and get on $[t_{\tilde N},t_{\tilde N+1})$ the inequality

$$\|Au(t)\| + \|A^{1-\rho}u'(t)\|$$

$$\le c(\|Au(t_{\tilde N})\| + (t-t_{\tilde N})^{-(1-\rho)}\|Au(t_{\tilde N})\|) +$$

$$+ \int_{t_{\tilde N}}^{t} \|e^{-(t-s)A}M'(s,u(s))\|\,ds +$$

$$+ \int_{t_{\tilde N}}^{t} \|A^{1-\rho}e^{-(t-s)A}M'(s,u(s))\|\,ds +$$

$$+ c\|M(t_{\tilde N},u(t_{\tilde N}))\| + c(t-t_{\tilde N})^{-(1-\rho)}\|M(t_{\tilde N},u(t_{\tilde N}))\| +$$

$$+ \|M(t,u(t))\|,$$

if $\tilde N \geq 2$. Using condition (V.1.5) we get

$$\int_{t_N}^{t} \|e^{-(t-s)A}M'(s,u(s))\| ds \leq c \int_{t_N}^{t} (\|Au(s)\|+1)^N \|A^{1-\rho}u'(s)\| \, ds,$$

$$\int_{t_{\tilde N}}^{t} \|A^{1-\rho}e^{-(t-s)A}M'(s,u(s))\| \, ds$$

$$\leq c \int_{t_{\tilde N}}^{t} \frac{1}{(t-s)^{1-\rho}} (\|Au(s)\|+1)^N \|A^{1-\rho}u'(s)\| \, ds,$$

and by condition (V.1.4) we obtain

$$\|M(t,u(t))\| \leq \|M(t,u(t))-M(t_{\tilde N},u(t_{\tilde N}))\| + \|M(t_{\tilde N},u(t_{\tilde N}))\|$$

$$\leq \|A(u(t)-u(t_N))\| \cdot g_1(\|u(t)-u(t_{\tilde N})\|_V + |t-t_{\tilde N}|)\hat k_D$$

$$+ \varepsilon'(\|A(u(t)-u(t_{\tilde N}))\|+1) \cdot g_2(t_{\tilde N}, \|Au(t_{\tilde N})\| + \|u(t)\|_V + \|u(t_{\tilde N})\|_V)$$

$$+ c(\varepsilon')g_2(t_{\tilde N}, \|Au(t_{\tilde N})\| + \|u(t)\|_V + \|u(t_{\tilde N})\|_V) +$$

$$+ k(T, \|u(t_{\tilde N})\|_V)\|Au(t_{\tilde N})\| .$$

If necessary δ can be diminished again, its smallness not depending $t_{\tilde N}$, in order to get

$$\|Au(t)\| + \|A^{1-\rho}u'(t)\|$$

$$\leq c(T,D,\|Au(t_{\tilde N})\|)\frac{1}{(t-t_{\tilde N})^{1-\rho}} +$$

$$+ c \int_{t_{\tilde N}}^{t} \frac{1}{(t-s)^{1-\rho}} (\|Au(s)\|+1)^N (\|Au(s)\| + \|A^{1-\rho}u'(s)\|) \, ds.$$

Using the lemma V.1.1 to follow this proof and our estimate for (V.1.9), we obtain for $q > \max(2, N/\rho)$ the estimate

$$\|Au(t)\| + \|A^{1-\rho}u'(t)\|$$

$$\leq C(T, D, \sup_{0 \leq t \leq t_{\tilde{N}}} \|Au(t)\|) \frac{1}{(t - t_{\tilde{N}})^{1-\rho}},$$

$t_{\tilde{N}} < t \leq t_{\tilde{N}+1}$. We can assume that $\sup_{0 = t_1 \leq t \leq t_2} \|Au(t)\|$ has been estimated if $\delta \leq \delta(\|A\varphi\|)$. This follows from the construction of the local strong solution in the proof of theorem II.1.1 . Thus proceeding stepwise we see that $\|Au(t)\|$ remains bounded if $t \uparrow T(\varphi)$. Therefore, according to theorem II.1.1, $[0, T(\varphi))$ is not the maximal interval of existence. This is a contradiction. The case $T(\varphi) = T$ may be treated analogously. Our theorem is proved. □

The lemma we have needed in the proof of the preceding theorem is as follows:

<u>Lemma V.1.1:</u> <u>Let</u> $T > 0$. <u>There is one and only one continuous solution</u> $y: (0, T] \to \mathbb{R}$ <u>of the integral equation</u>

$$y(t) = \frac{c_1}{t^{1-\rho}} + \int_0^t \frac{c_2}{(t-s)^{1-\rho}} h(s) y(s)\, ds$$

<u>with</u> $|y(t)| \leq \frac{c(c_1, c_2, \|h\|_{L^q((0,T))}, T)}{t^{1-\rho}}$, $0 < t \leq T < \infty$; <u>here</u> h <u>is a fixed function with</u> $h \in L^q((0, \tilde{T}))$, $0 < \tilde{T} < \infty$, <u>for some</u> $q > 1/\rho$.

<u>Proof:</u> The proof is based on iteration. Set $y_1(t) = \frac{c_1}{t^{1-\rho}}$,

$$y_2(t) = \frac{c_1}{t^{1-\rho}} + \int_0^t \frac{c_2}{(t-s)^{1-\rho}} h(s) \frac{c_1}{s^{1-\rho}}\, ds,$$

$$y_n(t) = \frac{c_1}{t^{1-\rho}} + \int_0^t \frac{c_2}{(t-s)^{1-\rho}} h(s) y_{n-1}(s)\, ds.$$

Then $|y_2(t)| \leq \frac{c_1}{t^{1-\rho}} + \frac{c_2 c_1}{t^{1-2\rho}} \int_0^1 \frac{1}{(1-\sigma)^{1-\rho}} \cdot \frac{1}{\sigma^{1-\rho}} h(t\sigma)\, d\sigma \leq$

$$\leq \frac{c_1}{t^{1-\rho}} + \frac{c_2 c_1}{t^{1-2\rho}} \left(\int_0^1 \frac{1}{(1-\sigma)^{\lambda(1-\rho)}\sigma^{\lambda(1-\rho)}}\, d\sigma\right)^{\frac{1}{\lambda}} \left(\int_0^1 |h(t\sigma)|^{\frac{\lambda}{\lambda-1}}\, d\sigma\right)^{\frac{\lambda-1}{\lambda}} \leq$$

$$\leq \frac{c_1}{t^{1-\rho}} + \frac{c_2 c_1}{t^{1-2\rho+\frac{\lambda-1}{\lambda}}} \|h\|_{L^{\lambda/(\lambda-1)}((0,t))} \frac{\Gamma(\lambda\rho-\lambda+1)^{2/\lambda}}{(\Gamma(2(\lambda\rho-\lambda+1)))^{1/\lambda}};$$ here we have chosen $\lambda/(\lambda-1) = q > 1/\rho$. This means that $0 < (\lambda-1)/\lambda < \rho$ and $1-2\rho+(\lambda-1)/\lambda = 1-\rho-(\rho-(\lambda-1)/\lambda) < 1-\rho$. By induction we get

$$|y_n(t)| \leq \frac{c_1}{t^{1-\rho}} \sum_{\nu=1}^{\infty} \frac{c_2^{\nu-1} \|h\|^{\nu-1}_{L^{\lambda/(\lambda-1)}((0,t))} \Gamma(\lambda\rho-\lambda+1)^{\nu/\lambda}}{\Gamma(\nu(\lambda\rho-\lambda+1))^{1/\lambda}} \cdot t^{(\nu-1)(\rho-(\lambda-1)/\lambda)},$$

$$|y_{n+1}(t)-y_n(t)| \leq \frac{c_1}{t^{1-\rho}} \frac{c_2^n \|h\|^n_{L^{\lambda/(\lambda-1)}((0,t))} \Gamma(\lambda\rho-\lambda+1)^{(n+1)/\lambda}}{\Gamma((n+1)(\lambda\rho-\lambda+1))^{1/\lambda}} \cdot t^{n(\rho-(\lambda-1)/\lambda)}.$$

Thus the functions y_n are uniformly convergent on every closed interval $I \subset (0,T]$; we set

$$y(t) = \lim_{n\to\infty} y_n(t), \quad t > 0.$$

Thus y is continuous on $(0,T]$, and because of our a-priori-estimate for $|y_n(t)|$ we see that $|y(t)|$ can be estimated as indicated in the lemma. Lebesgue's theorem shows that y fulfills our integral equation. If $c_1 = 0$ then

$$|t^{1-\rho}y(t)| \leq t^{\rho-(\lambda-1)/\lambda}\|h\|_{L^{\lambda(\lambda-1)}((0,t))}c_2 \cdot$$

$$\cdot c(0,c_2,\|h\|_{L^q((0,T))},T)\frac{\Gamma(\lambda\rho-\lambda+1)^2}{\Gamma(2(\lambda\rho-\lambda+1))}\cdot$$

$$\cdot \sup_{0<s<t} |s^{1-\rho}y(s)|, \quad 0 < t \leq T < \infty.$$

Since $\rho-(\lambda-1)/\lambda > 0$ we immediately get: $y(t) = 0$, $T \geq t > 0$. Thus the lemma is proved. □

§ 2. Applications to Parabolic Systems and to the Equations of Navier-Stokes

As for the examples they shall show how our abstract theorem works. First we deal with a nonlinear parabolic system; we intend to reduce the technicalities as much as possible in order to exhibit the decisive part in the application of theorem V.1.1. Secondly we interprete theorem V.1.1 as a regularity theorem for linear equations with "bad" coefficients as it is often possible with theorems for nonlinear equations; thus we give a second proof for the regularity of any weak solution of the Navier-Stokes equations which is in $C^0([0,T],(L^n(\Omega))^n)$.

As for parabolic systems we deal with the Cauchy problem, i.e. $\Omega = \mathbb{R}^n$. Our assumptions are as follows: Let $L \in \mathbb{N}$, $m \in \mathbb{N}$. We set

$$A = (-\Delta)^m + I;$$

we consider this operator in $L^2(\mathbb{R}^n)$ with domain of definition $D(A) = H^{2m,2}(\mathbb{R}^n)$.

Using the Fourier transform we immediately get the following results: A is a positive selfadjoint operator with

(V.2.1) $\frac{1}{c}\|u\|_{H^{2m,2}(\mathbb{R}^n)} \leq \|Au\|_{L^2(\mathbb{R}^n)} \leq c\|u\|_{H^{2m,2}(\mathbb{R}^n)}$,

$(\lambda+A)^{-1}$ exists as a bounded everywhere defined operator in $L^2(\mathbb{R}^n)$ and fulfills the estimate

(V.2.2) $\|(\lambda+A)^{-1}u\|_{L^2(\mathbb{R}^n)} \leq \frac{\hat{M}}{|\lambda|+1}\|u\|_{L^2(\mathbb{R}^n)}$, $\operatorname{Re}\lambda \geq 0$;

thus $-A$ generates an analytic semigroup e^{-tA} in $L^2(\mathbb{R}^n)$ with

(V.2.3) $\|e^{-tA}u\|_{L^2(\mathbb{R}^n)} \leq \hat{M}e^{-\delta t}$.

Here $\hat{M}$, δ are some positive constants. As for the powers A^α, $\alpha \geq 0$, we get

(V.2.4) $D(A^\alpha) = H^{2\alpha m,2}(\mathbb{R}^n)$,

(V.2.5) $\frac{1}{c(\alpha)}\|u\|_{H^{2\alpha m,2}(\mathbb{R}^n)} \leq \|A^\alpha u\|_{L^2(\mathbb{R}^n)} \leq c(\alpha)\|u\|_{H^{2\alpha m,2}(\mathbb{R}^n)}$.

We assume the space dimension

(V.2.6) $n > 2(m+1)$ for $m \geq 2$, and $n \leq 4$ for $m = 1$;

as it is seen from the proof of theorem V.2.1 below the case $1 \leq n \leq 2(m+1)$ may be considered as the trivial one. Afterwards we will see that the conclusion of theorem V.2.1 also holds if $n > 4$, $m = 1$. As for the nonlinearity let $f = (f^1,\dots,f^L)$; we assume that

(V.2.7) $f: \mathbb{R}^L \to \mathbb{R}^L$

is a C^1 function satisfying the following growth conditions: Let $q^* = \frac{n+2(m-1)}{n-2(m+1)}$ if $n > 2(m+1)$ and $1 < q^* < +\infty$ arbitrary if $m = 1$, $n \leq 4$,

$$\text{(V.2.8)} \quad |f(u)| \leq c(|u|^{q^*} + |u|), \quad u \in \mathbb{R}^L,$$

$$\text{(V.2.9)} \quad \left|\frac{\partial f^l}{\partial u^k}(u)\right| \leq c|u|^{q^*-1}, \quad u = (u^1, \dots, u^L) \in \mathbb{R}^L,$$

$$\text{(V.2.10)} \quad \sum_{1 \leq l,k \leq L} \frac{\partial f^l}{\partial u^k}(u)\xi^l \xi^k \geq -c|\xi|^2, \quad u \in \mathbb{R}^L, \ \xi \in \mathbb{R}^L$$

with some $c > 0$. Then the following theorem holds:

<u>Theorem V.2.1:</u> <u>Let</u> φ <u>be a real valued function with</u>

$$\varphi \in D(A).$$

<u>If we set</u> $M(u) = f(u)$, $u \in (D(A))^L$ <u>then</u> M <u>is a mapping from</u> $D((A^{1-\rho})^L)$ <u>into</u> $(L^2(\mathbb{R}^n))^L$ <u>for some</u> $\rho \in (0,1)$; <u>moreover</u>, <u>for some</u> <u>continuous function</u> $k: \mathbb{R}^+ \to \mathbb{R}^+$

$$\text{(V.2.11)} \quad \|M(u) - M(v)\|_{(L^2(\mathbb{R}^n))^L}$$

$$\leq k(\|A^{1-\rho}u\|_{(L^2(\mathbb{R}^n))^L} + \|A^{1-\rho}v\|_{(L^2(\mathbb{R}^n))^L}) \cdot$$

$$\cdot \|A^{1-\rho}(u-v)\|_{(L^2(\mathbb{R}^n))^L}.$$

<u>Therefore there exists a</u> $T(\varphi) \in (0, +\infty]$ <u>such that there is a</u> <u>unique real valued</u> $u \in C^1([0,T(\varphi)), (L^2(\mathbb{R}^n))^L)$ <u>with</u> $u(t) \in D(A)$, $0 \leq t < T(\varphi)$, $Au \in C^0([0,T(\varphi)), (L^2(\mathbb{R}^n))^L)$

$$u' + Au + M(u) = 0, \quad 0 \leq t < T(\varphi),$$

$$u(0) = \varphi,$$

$$\lim_{t \uparrow T(\varphi)} \|Au(t)\| = +\infty \text{ if } T(\varphi) < +\infty.$$

Moreover under our assumptions above on f we have

(V.2.12) $T(\varphi) = +\infty.$

Proof: As for the first part of the preceding theorem we have the estimate

$$\| f(u)-f(v) \|_{(L^2(\mathbb{R}^n))^L}$$

$$\leq c(\| \, |u|^{\frac{4m}{n-2(m+1)}}+|v|^{\frac{4m}{n-2(m+1)}} \|_{L^{2q_1}(\mathbb{R}^n)} +1)\| u-v \|_{(L^{2q_2}(\mathbb{R}^n))^L}$$

with

$$\frac{1}{2q_2} = \frac{1}{2} - \frac{2m(1-\rho)}{n},$$

$$\frac{1}{2q_1} = \frac{2m(1-\rho)}{n};$$

choosing ρ sufficiently small in (0,1) we get

$$\frac{n-2(m+1)}{4m} \, \frac{2m(1-\rho)}{n} = \frac{1-\rho}{2n}(n-2(m+1)) = (\frac{1}{2} - \frac{m+1}{n})(1-\rho)$$

$$\geq \frac{1}{2} - \frac{2m(1-\rho)}{n}$$

for $m \geq 2$. Using the continuous imbedding $H^{2m(1-\rho),2}(\mathbb{R}^n) \subset L^q(\mathbb{R}^n)$, $\frac{1}{q} \geq \frac{1}{2} - \frac{2m(1-\rho)}{n}$, $q \geq 2$ we arrive at (V.2.11) in the case $m \geq 2$. If $m = 1$, $n \leq 4$ we can use the continuous imbedding

$$H^{2m(1-\rho),2}(\mathbb{R}^n) \subset L^q(\mathbb{R}^n)$$

for any $q \geq 2$ if ρ is small enough. Set $B = (L^2(\mathbb{R}^n))^L$. (V.2.4) together with theorem II.1.1 now furnishes the first part of our theorem; that u is real valued immediately follows from the construction of u given in the proof of theorem II.1.1 . As for the second part let $m \geq 2$; we observe that by the Gagliardo-Nirenberg interpolatory inequality we get

$$\|M(u)\|_{(L^2(\mathbb{R}^n))^L} \leq c\|u\|_{(H^{2m,2}(\mathbb{R}^n))^L} \cdot (\|u\|_{(H^{m+1,2}(\mathbb{R}^n))^L} + 1)^{\frac{4m}{n-2(m+1)}},$$

$$\leq c\|Au\|_{(L^2(\mathbb{R}^n))^L} \cdot (\|u\|_{(H^{m+1,2}(\mathbb{R}^n))^L} + 1)^{\frac{4m}{n-2(m+1)}},$$

$$\|M(u)-M(v)\|_{(L^2(\mathbb{R}^n))^L} \leq c\| \, |u|^{\frac{n+2(m-1)}{n-2(m+1)}} + |u| + |v|^{\frac{n+2(m-1)}{n-2(m+1)}} + |v| \, \|_{L^2(\mathbb{R}^n)}$$

$$\leq c\| \, |u-v|^{\frac{n+2(m-1)}{n-2(m+1)}} + |u-v| \, \|_{L^2(\mathbb{R}^n)} + c\| \, |v|^{\frac{n+2(m-1)}{n-2(m+1)}} + |v| \, \|_{L^2(\mathbb{R}^n)},$$

$$\leq c\|A(u-v)\|_{(L^2(\mathbb{R}^n))^L} \|u-v\|^{\frac{4m}{n-2(m+1)}}_{(H^{m+1,2}(\mathbb{R}^n))^L} + c\|u-v\|_{(L^2(\mathbb{R}^n))^L} (1 + \|Av\|^{1+\frac{4m}{n-2(m+1)}}_{(L^2(\mathbb{R}^n))^L} + \|v\|_{(L^2(\mathbb{R}^n))^L}),$$

$$u, v \in D(A);$$

here we have used the Gagliardo-Nirenberg inequality in the form

$$\|u\|_{L^{2\frac{n+2(m-1)}{n-2(m+1)}}(\mathbb{R}^n)} \leq c\|u\|_{H^{2m,2}(\mathbb{R}^n)}^{\frac{n-2(m+1)}{n+2(m-1)}} \cdot \|u\|_{L^{\hat{q}}(\mathbb{R}^n)}^{1-\frac{n-2(m+1)}{n+2(m-1)}},$$

$u \in H^{2m,2}(\mathbb{R}^n)$, $\frac{1}{\hat{q}} = \frac{1}{2} - \frac{m+1}{n}$,

$$\|u\|_{L^{\hat{q}}(\mathbb{R}^n)} \leq c\|u\|_{H^{m+1,2}(\mathbb{R}^n)}.$$

Setting $\hat{k}(T,x) = \hat{k}(x) = c(x+1)^{4m/(n-2(m+1))}$, $g_1(x) = cx^{4m/(n-2(m+1))}$, $g_2(s,x) = g_2(x) = (1+x)^{4m/(n-2(m+1))}$, $x \geq 0$, we see that assumptions (V.1.2), (V.1.3) and (V.1.4) are fulfilled with $V = H^{m+1,2}(\mathbb{R}^n)$. The case $m = 1$, $n \leq 4$, is the easy one and is left to the reader. As for assumption (V.1.6) we get by scalar multiplication of $u' + Au = \tilde{f}$ with Au the inequality

$$\text{(V.2.13)} \quad \int_0^T \|u'(t)\|^2_{(L^2(\mathbb{R}^n))^L} dt + \sup_{0\leq t\leq T} \|A^{\frac{1}{2}}u(t)\|^2_{(L^2(\mathbb{R}^n))^L} + \int_0^T \|Au(t)\|^2_{(L^2(\mathbb{R}^n))^L}$$

$$\leq c(\|A^{\frac{1}{2}}u(0)\|^2_{(L^2(\mathbb{R}^n))^L} + \int_0^T \|\tilde{f}(t)\|^2_{(L^2(\mathbb{R}^n))^L} dt)$$

for $T > 0$ with a constant c not depending on T; here $f \in L^2((0,T), (L^2(\mathbb{R}^n))^L)$, $u' \in L^2((0,T),(L^2(\mathbb{R}^n))^L)$, $u(t) \in D(A)$ a.e. in $(0,T)$, $Au \in L^2((0,T),(H^{2m,2}(\mathbb{R}^n))^L)$, from which in particular follows: $A^{1/2}u \in C^0([0,T],(L^2(\mathbb{R}^n))^L)$. This is a consequence of the self-adjointness of A. The reader may easily prove this by himself using (V.2.13) first for $f_{\hat{m}}, u_{\hat{m}}$ instead of f,u with $f_{\hat{m}}, u_{\hat{m}}$ being somewhat more regular than f,u, and then approximating f,u by $f_{\hat{m}}, u_{\hat{m}}$; we have also used an analogous result for the Navier-

Stokes equations; an abstract proof in a more general situation with respect to A can be found in [W11,I]. It is also a consequence of [W 11,II] that (V.2.13) implies for $u(0) \in D(A)$

$$\text{(V.2.14)} \quad \int_0^T \|u'(t)\|^q_{(L^2(\mathbb{R}^n))^L}\,dt + \int_0^T \|Au(t)\|^q_{(L^2(\mathbb{R}^n))^L}\,dt$$

$$\leq c(q,T)\,(\|A\varphi\|^q_{(L^2(\mathbb{R}^n))^L} + \int_0^T \|\tilde{f}(t)\|^q_{(L^2(\mathbb{R}^n))^L}\,dt),$$

$q \geq 2$, $T > 0$ with a constant $c(q,T)$ possibly depending on T and being monotonically non decreasing with respect to T; of course we must assume that $\tilde{f} \in L^q((0,T),(L^2(\mathbb{R}^n))^L)$ and that u has corresponding regularity properties (i.e. replace 2 by q in the previous assumptions on u). The reader may verify that (V.2.13) can also be proved by consideration of

$$\hat{u}'(t,\xi) + (|\xi|^{2m}+1)\hat{u}(t,\xi) = \hat{f}(t,\xi),$$

$t \geq 0$, $\xi \in \mathbb{R}^n$, where $\hat{u}$ is the Fourier transform of u with respect to the space variables. Thus also assumption (V.1.6) is fulfilled for any $T > 0$. Now we assume that $T(\varphi) < +\infty$. Let $T < T(\varphi)$. As for the solution u of $u' + Au + M(u) = 0$ on $[0,T(\varphi))$ having already been constructed in the first part of theorem V.2.1 it follows from theorem II.1.1 that $u'(t) \in D(A^{1-\rho})$, $0 < t < T(\varphi)$,

$$A^{1-\rho}u' \in C^0((0,T(\varphi)),(L^2(\mathbb{R}^n))^L),$$

and consequently

$$Au + f(u) \in C^0((0,T(\varphi)),(H^{1,2}(\mathbb{R}^n))^L)$$

since we can choose ρ sufficiently small. The putative derivatives $\frac{\partial}{\partial x_i}f^1(u)$ can easily be proved to be in $C^0([0,T(\varphi)),$ $(L^2(\mathbb{R}^n))^L)$ with the aid of (V.2.9), Hölder's inequality and the Sobolev imbedding theorem of our auxiliary properties since $u \in C^0([0,T(\varphi)),(H^{2m,2}(\mathbb{R}^n))^L)$. Thus

$$Au \in C^{0}((0,T(\varphi)),(H^{1,2}(\mathbb{R}^{n}))^{L}).$$

Using the Fourier transform we get

$$u \in C^{0}((0,T(\varphi)),H^{2m+1,2}(\mathbb{R}^{n})).$$

We have

$$\frac{1}{h}\{u'(t+h)-u'(t)+A(u(t+h)-u(t))+f(u(t+h))-f(u(t))\} = 0,$$

$$0 \leq t < t+h < T(\varphi).$$

Scalar multiplication of this equation by $\frac{1}{h}(u(t+h)-u(t))$ gives

$$\text{(V.2.15)} \quad \frac{d}{dt}\left\|\frac{u(t+h)-u(t)}{h}\right\|^{2}_{(L^{2}(\mathbb{R}^{n}))^{L}} + 2\left\|A^{\frac{1}{2}}\frac{u(t+h)-u(t)}{h}\right\|^{2}_{(L^{2}(\mathbb{R}^{n}))^{L}}$$

$$+ 2\left(\frac{f(u(t+h))-f(u(t))}{h},\frac{u(t+h)-u(t)}{h}\right) = 0.$$

Since

$$f^{l}(u(t+h)) - f^{l}(u(t)) = \sum_{K=1}^{L}\int_{0}^{1}\frac{\partial f^{l}}{\partial u^{k}}(\tau u(t+h)+(1-\tau)u(t))\,d\tau$$

$$(u^{k}(t+h)-u^{k}(t))$$

we get with (V.2.10) the inequality

$$2\left(\frac{f(u(t+h))-f(u(t))}{h},\frac{u(t+h)-u(t)}{h}\right)$$

$$\geq -2c\left\|\frac{u(t+h)-u(t)}{h}\right\|^{2}_{(L^{2}(\mathbb{R}^{n}))^{L}}.$$

By integration we obtain

$$\int_{0}^{T}\left\|A^{\frac{1}{2}}\frac{u(t+h)-u(t)}{h}\right\|^{2}_{(L^{2}(\mathbb{R}^{n}))^{L}}dt$$

$$\leq T\left\|\frac{u(h)-\varphi}{h}\right\|^{2}_{(L^{2}(\mathbb{R}^{n}))^{L}}e^{2cT},\quad T < T(\varphi).$$

Letting h tend to 0 we get

$$\int_0^T \|A^{\frac{1}{2}}u'(t)\|^2\,dt \le T\|A\varphi+f(\varphi)\|^2_{L^2(\mathbb{R}^n)} e^{2cT}.$$

In particular

$$u' \in L^2((0,T(\varphi)),(H^{1,2}(\mathbb{R}^n))^L),$$

$$Au_{x_i} + f'(u)u_{x_i} = -u'_{x_i},\quad 1 \le i \le n,$$

where $f'(u)v$ is the vector with components $\sum_{k=1}^{L} \frac{\partial f^l}{\partial u^k}(u)v^k$. Now we consider the elliptic system

$$\text{(V.2.16)}\quad Aw + (c+1)w + f'(u(t))w = \sigma(-u'_{x_i} + (c+1)u)(t) = \sigma F(t),$$

$0 \le \sigma \le 1$. Let $w_{\sigma_1}, w_{\sigma_2} \in (H^{2m,2}(\mathbb{R}^n))^L$ be two solutions of (V.2.15) belonging to the parameter values σ_1, σ_2. Then

$$\|w_{\sigma_1} - w_{\sigma_2}\|_{(H^{m,2}(\mathbb{R}^n))^L} \le c|\sigma_2-\sigma_1|\ \|F(t)\|_{(L^2(\mathbb{R}^n))^L};$$

observe that for $n > 4m$, $m \ge 2$

$$\text{(V.2.17)}\quad \|f'(u(t))w\|_{(L^2(\mathbb{R}^n))^L} \le$$

$$\le c[(\int_{|x|\le R} |u|^{\frac{8m}{n-2(m+1)}}|w|^2\,dx)^{\frac{1}{2}} + (\int_{|x|\ge R} |u|^{\frac{8m}{n-2(m+1)}}|w|^2\,dx)^{\frac{1}{2}}]$$

$$\le c[(\int_{|x|\le R} |u|^{\frac{8m}{n-2(m+1)}p_1}dx)^{\frac{1}{2p_1}}(\int_{|x|\le R} |w|^{2q_1}dx)^{\frac{1}{2q_1}}$$

$$+ \left(\int_{|x|\geq R} |u|^{\frac{8m}{n-2(m+1)}p_2}dx\right)^{\frac{1}{2p_2}} \left(\int_{|x|\geq R} |w|^{2q_2}dx\right)^{\frac{1}{2q_2}}]$$

with $\frac{1}{2p_2} = \frac{2m}{n}$, $\frac{1}{2q_2} = \frac{1}{2}-\frac{2m}{n}$, $\frac{1}{2p_1} = \frac{(n-4m)4m}{2n(n-2(m+1))} < \frac{2m}{n}$, $\frac{1}{2q_1} = \frac{1}{2}-\frac{1}{2p_1} > \frac{1}{2}-\frac{2m}{n}$. Finally we get for s with $\frac{1}{2q_1} = \frac{1}{2}-\frac{s}{n}$

$$\text{(V.2.18)} \quad \|f'(u(t))w\|_{(L^2(\mathbb{R}^n))^L}$$

$$\leq c[\|u(t)\|^{\frac{4m}{n-2(m+1)}}_{(H^{2m,2}(|x|\leq R))^L} \cdot \|w\|_{H^{s,2}(|x|\leq R)^L} +$$

$$+ \|u(t)\|^{\frac{4m}{n-2(m+1)}}_{(H^{m+1,2}(|x|\geq R))^L} \cdot \|w\|_{(H^{2m,2}(|x|\geq R))^L}].$$

If $\tilde{w}_\sigma = T_\sigma w$ is the unique solution of

$$A\tilde{w}_\sigma + (c+1)\tilde{w}_\sigma + f'(u(t))w = \sigma F(t)$$

we thus have shown that the mapping T_σ is compact from $(H^{2m,2}(\mathbb{R}^n))^L$ into itself. The term $\|f'(u(t))w\|_{(L^2(\mathbb{R}^n))^L}$ can also be estimated in the following way ($q^* = \frac{n+2(m-1)}{n-2(m+1)}$)

$$\text{(V.2.19)} \quad \|f'(u(t))w\|_{(L^2(\mathbb{R}^n))^L} \leq c\|u\|^{\frac{4m}{n-2(m+1)}}_{(L^{2q^*}(\mathbb{R}^n))^L}\|w\|_{(L^{2q^*}(\mathbb{R}^n))^L}$$

$$\leq c\|Au(t)\|^{\alpha\frac{4m}{n-2(m+1)}}_{(L^2(\mathbb{R}^n))^L}\|u\|^{(1-\alpha)\frac{4m}{n-2(m+1)}}_{(H^{m+1,2}(\mathbb{R}^n))^L} \cdot$$

$$\cdot \|Aw\|^{\alpha}_{(L^2(\mathbb{R}^n))^L}\|w\|^{1-\alpha}_{(H^{m+1,2}(\mathbb{R}^n))^L}$$

$$\leq c(\varepsilon)\|Aw\|_{(L^2(\mathbb{R}^n))^L}\|w\|^{4m/(n-2(m+1))}_{(H^{m+1,2}(\mathbb{R}^n))^L} +$$

$$+ \varepsilon\|Au(t)\|_{(L^2(\mathbb{R}^n))^L}\|u(t)\|^{4m/(n-2(m+1))}_{(H^{m+1,2}(\mathbb{R}^n))^L},$$

$\varepsilon > 0$, $\alpha = (\frac{n+2(m-1)}{n-2(m+1)})^{-1}$. Let $w_{\sigma_1} \in (H^{2m,2}(\mathbb{R}^n))^L$ be the unique solution of (V.2.16) belonging to the parameter value σ_1 of σ. Considering the equation

$$A(w_{\sigma_2}-w_{\sigma_1}) + (c+1)(w_{\sigma_2}-w_{\sigma_1}) + f'(u(t))(w_{\sigma_2}-w_{\sigma_1})$$

$$= (\sigma_2-\sigma_1)F(t)$$

and applying (V.2.19) with $w_{\sigma_2}-w_{\sigma_1}$ instead of w to the inequality

$$\|A(w_{\sigma_2}-w_{\sigma_1})\|_{(L^2(\mathbb{R}^n))^L} \leq (c+1)\|w_{\sigma_2}-w_{\sigma_1}\|_{(L^2(\mathbb{R}^n))^L}$$

$$+ \|f'(u(t))(w_{\sigma_2}-w_{\sigma_1})\|_{(L^2(\mathbb{R}^n))^L}$$

$$+ |\sigma_2-\sigma_1|\ \|F(t)\|_{(L^2(\mathbb{R}^n))^L}$$

we get the a-priori estimate

$$\text{(V.2.19)}\quad \|A(w_{\sigma_2}-w_{\sigma_1})\|_{(L^2(\mathbb{R}^n))^L}$$

$$\leq c(\|F(t)\|_{(L^2(\mathbb{R}^n))^L} + \varepsilon\|Au(t)\|_{(L^2(\mathbb{R}^n))^L}\cdot$$

$$\cdot\|u(t)\|^{4m/(n-2(m+1))}_{(H^{m+1,2}(\mathbb{R}^n))^L})$$

if $\sigma_1 \leq \sigma_2 \leq \delta(\varepsilon)+\sigma_1$. $\delta(\varepsilon)$ is a sufficiently small positive quantity but its smallness neither depends on $u(t)$ nor on σ_1. If we assume that for $\sigma \in [0,\sigma_1]$ the solution w_σ exists and can be estimated a priori then the same is true for $\sigma \in [0,\sigma_1+\delta(\varepsilon)]$: (V.2.19) gives the a priori estimate; this estimate of course also holds for all fixed points of τT_σ, $0 \le \tau \le 1$, $0 \le \sigma \le \sigma_1 + \delta(\varepsilon)$. Since T_σ is compact we also get the existence of w_σ. Starting with $\sigma = 0$ we can thus proceed stepwise; after finitely many steps we reach $\sigma = 1$ and the estimate $(w = w_1)$:

$$\|Aw\|_{(L^2(\mathbb{R}^n))^L} \le c(\varepsilon)\|F(t)\|_{L^2(\mathbb{R}^n)^L} + \varepsilon\|Au(t)\|_{(L^2(\mathbb{R}^n))^L} \cdot$$

$$\cdot \|u(t)\|^{4m/(n-2(m+1))}_{(H^{m+1,2}(\mathbb{R}^n))^L}$$

where we wish to estimate $\|u(t)\|_{(H^{m+1,2}(\mathbb{R}^n))^L}$ a priori. (V.2.15) gives the a priori estimate

$$\text{(V.2.21)} \quad \|u'(t)\|_{(L^2(\mathbb{R}^n))^L} \le c, \quad 0 \le t < T(\varphi),$$

where c does not depend on t. We also have estimated a priori

$$\int_0^{T(\varphi)} \|A^{\frac{1}{2}}u'(t)\|^2_{(L^2(\mathbb{R}^n))^L}\, dt;$$

thus

$$\text{(V.2.22)} \quad \|u(t)\|_{(H^{m,2}(\mathbb{R}^n))^L} \le c, \quad 0 \le t < T(\varphi).$$

Multiplying the equation

$$u'_{x_i} + Au_{x_i} + f'(u)u_{x_i} = 0, \quad 0 < t < T(\varphi)$$

scalarly by u_{x_i}, using $(u'_{x_i}, u_{x_i}) = -(u', u_{x_i x_i})$ and (V.2.22), (V.2.21) we get

$$\text{(V.2.23)} \quad \|u(t)\|_{(H^{m+1,2}(\mathbb{R}^n))^L} \le c.$$

Observe that $w = w_1 = u_{x_i}(t)$. Choosing ε sufficiently small (its smallness only depending on imbedding constants) we get

(V.2.24) $\|Au_{x_i}(t)\|_{(L^2(\mathbb{R}^n))^L} \le c(\varepsilon)\|F(t)\|_{(L^2(\mathbb{R}^n))^L}$,

where $c(\varepsilon)$ does not depend on t. Together with $\int_0^{T(\varphi)} \|A^{\frac{1}{2}}u'(t)\|^2 dt < \infty$ we get

$$u \in C^0([0,T(\varphi)],(H^{m+1,2}(\mathbb{R}^n))^L)$$

and our theorem for $n > 4m$, $m \ge 2$. The remaining cases can be treated analogously and may be left to the reader. □

We want to make some important remarks in this context:
1. The elliptic operator must not be the same in each row of our system $u' + Au + M(u) = 0$, i.e. instead of A we can allow $A^l = \gamma_l((-\Delta)^m+I)$ with some positive constant γ_l, $1 \le l \le L$. More general we may have

$$A^l(t) = \sum_{\substack{|\alpha|\le m,\\ |\beta|\le m}} D^\alpha(A^l_{\alpha\beta}(t,x)D^\beta),\ 1 \le l \le L,\ t \ge 0,$$

where the $A^l(t)$ are positive selfadjoint operators with sufficiently regular coefficients and domain of definition $D(A^l(t)) = D(A^l(0)) = H^{2m,2}(\mathbb{R}^n)$. 2. The case $m = 1$ is only seemingly an exceptional one. Taking the evolution operator $U^l_p(t,s)$, $t \ge s \ge 0$, belonging to the $A^l(t)$ in $L^p(\mathbb{R}^n)$ for $p > n$ we can construct the local strong solution u in $B = (L^p(\mathbb{R}^n))^L$ with the result that $u \in C^0([0,T(\varphi)),C^{2m-1+\alpha}(\mathbb{R}^n))$ for some $\alpha \in (0,1)$; then the $A^l(t)$ have to be considered as unbounded operators in $L^p(\mathbb{R}^n)$ with domain of definition $D(A^l(t)) = H^{2m,p}(\mathbb{R}^n)$. The assumptions (V.1.2), (V.1.3), (V.1.4), (V.1.6) can be shown to be fulfilled; for (V.1.6) we have to take p sufficiently large. (V.1.5) is not needed (see the remark in V.1. after our assumptions I., II.). It turns out that the space V in (V.1.2), (V.1.3) and (V.1.4) is the same for all $p \ge 2$, namely $(H^{m+1,2}(\mathbb{R}^n))^L$. By this the calculations being necessary to show that $T(\varphi)=+\infty$ become somewhat simpler if $p>n$, and also the case $m=1$ is included. Moreover the global solution is in $C^{o}([0,+\infty),C^{2m-1+\alpha}(\mathbb{R}^{n}))$ Nevertheless we have not used this access since we did not want to exhibit the linear theory in the case of a general p. There is also a theorem like V.2.1 for the initial boundary value problem with boundary values O, but then $q^*=\frac{n+2m}{n-2m}$; the exponent $q^*=\frac{n+2(m-1)}{n-2(m+1)}$ may be reached in the case of a single equation and for space dimensions n being large enough (see [Goe]). □

Now we turn to the application of our abstract theorem to the Navier-Stokes equations; this application is a rather indirect one. Let $n\geq 3$. Let $p>n$, $T>0$. Let

$$\varphi\in(H^{2,p}(\Omega))^n\cap(\overset{o}{H}{}^{1,p}(\Omega))^n \text{ real, } \nabla\cdot\varphi=0$$

$$f\in C^{\frac{1}{2p}}([0,T],(L^p(\Omega))^n)\cap C^o((0,T],(H^{\frac{1}{p},p}(\Omega))^n) \text{ real,}$$

where $\Omega\subset\mathbb{R}^n$ is as in IV. Let u be a weak solution of the Navier-Stokes equations over $(0,T)\times\Omega$ with data φ,f in the sense of definition IV.1.1. Then the following regularity theorem holds:

<u>Theorem V.2.2:</u> <u>Let</u> u <u>be as just introduced</u>. <u>Let</u>

$$u\in C^o([0,T],(L^n(\Omega))^n).$$

<u>Then</u>

$$u\in C^o([0,T],(H^{2,p}(\Omega))^n\cap(\overset{o}{H}{}^{1,p}(\Omega))^n),$$

$$\nabla\cdot u(t)=0,\quad 0\leq t\leq T,$$

$$u\in C^1([0,T],(L^p(\Omega))^n),$$

$$u'\in\bigcap_{\substack{0<\varepsilon<T,\\ 0\leq s<2}} C^o([\varepsilon,T],(H^{s,p}(\Omega))^n).$$

<u>Proof:</u> We assume that the potential theoretical estimates of Solonnikov (Theorem III.1.1) hold for any $n \geq 2$, and consequently that $A = A_p = -P_p\Delta$ generates an analytic semigroup in $H_p(\Omega)$ for any n (cf. the section III.4). Following an idea of Gerhardt [G] we introduce the linear equation

$$\text{(V.2.25)} \quad \left\{ \begin{aligned} w' - \nu\Delta w + u_k\cdot\nabla w + \nabla\tilde{p} &= f, \\ \nabla\cdot w &= 0, \\ w(0) &= \tilde{\varphi}, \end{aligned} \right.$$

where $u_k \to u$ in $C^0([0,T],(L^n(\Omega))^n)$, $k\to\infty$, $u_k \in C^1([0,T]\times\bar{\Omega})$, $\nabla\cdot u_k = 0$, $\sup\limits_{\substack{|t-s|\leq r\\ t,s\in[0,T]}} \|u_k(t)-u_k(s)\|_{(L^n(\Omega))^n} \leq \sup\limits_{\substack{|t-s|\leq r\\ t,s\in[0,T]}} \|u(t)-u(s)\|_{(L^n(\Omega))^n} + \varepsilon(r)$ with $\varepsilon(r)\to 0$ for $r\to 0$, and where u is the weak solution under consideration. Let $2 < q < n$. We set

$$M_k(t,w) = P_q(u_k(t)\cdot\nabla w),$$
$$B = P_q(L^q(\Omega))^n.$$

The mapping M_k fulfills the following estimate:

$$\|M_k(t,w_2)-M_k(s,w_1)\|$$

$$\leq \|M_k(s,w_2)\| + \|M_k(s,w_1)\| + \\ + c(q,n,\Omega)\|u_k(t)-u_k(s)\|_{(L^n(\Omega))^n}\cdot\|w_2\|_{1,\frac{qn}{n-q}},$$

$$\leq \|M_k(s,w_2)\| + \|M_k(s,w_1)\| + \\ + c(q,n,\Omega)\|u_k(t)-u_k(s)\|_{(L^n(\Omega))^n}\|w_2\|_{2,q},$$

$$\leq c(q,n,\Omega)\|u_k(t)-u_k(s)\|_{(L^n(\Omega))^n}\|w_2-w_1\|_{2,q} +$$

$$+ c(q,n,\Omega)\|u_k(t)-u_k(s)\|_{(L^n(\Omega))^n}\|w_1\|_{2,q} + \|M_k(s,w_2)\| + \|M_k(s,w_1)\|.$$

Let $\hat{c} \geq \sup_{0\leq t\leq T} (\|u(t)\|+\|u_k(t)\|)$. Then we get

$$\|M_k(t,w_2)-M_k(s,w_1)\| \leq$$

$$\leq \|w_2-w_1\|_{2,q}\tilde{g}_1(|t-s|) + c(q,n,\hat{c},\Omega)\|w_1\|_{2,q} + \|M_k(s,w_2)\| + \|M_k(s,w_1)\|,$$

$$\leq \|w_2-w_1\|_{2,q}\tilde{g}_1(|t-s|) + c(q,n,\hat{c},\Omega)\|w_1\|_{2,q} + \|u_k(s)\|_{(C^o(\overline{\Omega}))^n}\|w_2-w_1\|_{1,q} + c(q,n,\Omega)\|u_k(s)\|_{(L^n(\Omega))^n}\|w_1\|_{2,q},$$

$$\leq \|w_2-w_1\|_{2,q}\tilde{g}_1(|t-s|) + c(q,n,\hat{c},\Omega)\|w_1\|_{2,q} + \|u_k(s)\|_{(C^o(\overline{\Omega}))^n}\|w_2-w_1\|_{1,q},$$

$$\leq \|w_2-w_1\|_{2,q}\tilde{g}_1(|t-s|) + c(q,n,\hat{c},\Omega)(\|w_1\|_{2,q}+\|u_k(s)\|_{(C^o(\overline{\Omega}))^n})(\|A_q^{\frac{1}{2}}(w_2-w_1)\|+1)$$

where $\tilde{g}_1(|t-s|) = c(q,n,\Omega)(\sup_{|s''-s'|\leq|t-s|} \|u(s'')-u(s')\|_{(L^n(\Omega))^n} + \varepsilon(|t-s|))$ and thus $\tilde{g}_1(r) \to 0 = \tilde{g}_1(0)$ for $r \to 0$.

Setting $V = B$ we see from the preceding inequalities that M_k fulfills the inequality (V.1.3) and the inequality (V.1.4).

Observe that the constants in the latter estimate do not depend on k with the exception of g_2. M_k fulfills (V.1.5) with $N=1$. Because of the regularity properties of u_k it is not difficult to show that M_k also fulfills the first one of the inequality (V.1.1) with $\rho=\frac{1}{2}$ (cf. theorem III.2.6). Thus theorem V.1.1 in connection with theorem III.1.1 for arbitrary n shows that the equation

$$\begin{aligned} w' + Aw + P_q M_k(t,w) &= P_q f, \\ w(0) &= \tilde{\varphi} \end{aligned}$$

has a unique solution over [O,T] in B with the properties listed in theorem V.1.1, second case. From this solution we get in the usual way a pair $(w,\tilde{p})$ which solves (V.2.25). Let us apply now (V.1.9) for $s=0=t_1$, $\tilde{N}=1$. This gives an a-priori estimate for $\|Aw(t)\|$ on $[0,\delta]$ which does not depend on k. Observe that δ also does not depend on k. Letting k tend to ∞ we get an element $w \in C^0([0,\delta],B)$ with

$$w' \in L^\infty((0,\delta),B),$$

$$Aw \in L^\infty((0,\delta),B),$$

$$\begin{aligned} w' + Aw + P_q(u\cdot\nabla w) &= P_q f, \\ w(0) &= \tilde{\varphi} \end{aligned}$$

on $(0,\delta)$. Defining the notion of a weak solution to

$$\text{(V.2.26)} \quad \left\{ \begin{aligned} w' - \nu\Delta w + u\cdot\nabla w + \nabla\tilde{p} &= f, \\ \nabla\cdot w &= 0, \\ w(0) &= \tilde{\varphi} \end{aligned} \right.$$

over Q_T analogously to (IV.1.1) we see that because of

$$(u\cdot\nabla w, w) = 0$$

a real weak solution to (V.2.26) could be constructed completely analogously to [Li,p. 64-77]; moreover we can show as in the proof of theorem IV.3.1 that

$$\operatorname*{ess\,sup}_{0<t<T} \|w(t)\|_{L^2(\Omega)} + \nu \int_0^T \| w(s)\|^2_{L^2(\Omega)} \, ds \le$$

$$\le \|\tilde{\varphi}\|_{L^2(\Omega)} + c(\nu,n,\Omega) \int_0^T \|f(s)\|^2 \, ds.$$

Thus w is unique and therefore

$$w(t) = u(t).$$

The element $w \in C^0([0,\delta],B)$ with $w' \in L^\infty((0,\delta),B)$, $Aw \in L^\infty((0,\delta),B)$ just constructed constitutes of course a weak solution to (V.2.26) over $Q_\delta = (0,\delta)\times\Omega$. Therefore

$$u(t) \in D(A) \text{ on } (0,\delta),$$
$$u \in L^\infty((0,\delta),H^{2,q}(\Omega)).$$

Choosing $q = \frac{n(n+1)}{2n} \in (1,n)$ we get that

$$\|A_q f\| \ge c\|f\|_{C^0(\bar{\Omega})}, \quad f \in D(A), \quad u \in L^\infty((0,\delta),L^{n+1}(\Omega)).$$

The regularity-theorem IV.5.1 (only a weaker version with $\frac{n}{r}+\frac{2}{s} < 1$, $r > n$, $s = +\infty$ is needed) gives then the desired result for $T = \delta$. Since δ did not depend on k, we can repeat the whole procedure on $(\delta,2\delta)$ with $w(\delta) = u(\delta)$ and so on. Thus theorem V.2.2 follows.

Remark: We see that it is rather the method of proof we have needed than theorem V.1.2 itself. In view of theorem III.1.1 the assumption $\varphi \in (W^{2-2/p,p}(\Omega))^n \cap (\overset{\circ}{H}{}^{1,p}(\Omega))^n$, $\nabla\cdot\varphi = 0$ is sufficient.

§ 3. Comments to chapter V.

We restrict ourselves to some remarks to V.1 and V.2 (mostly bibliographical).

<u>To § 1:</u> Most results for global (in t) regularity for abstract semilinear parabolic equations rest on the comparison of the linear part with the nonlinearity. It may be clear then that M(u) may grow at most like $\|Au\|$ with respect to the V-norm of u. V can be considered as the space where u can be estimated a priori. It is clear that the admissible growth of M(u) becomes stronger if the norm of V becomes stronger. An earlier version of theorem V.1.1 was published in [W 9].

<u>To § 2:</u> Theorems corresponding to V.2.2, but for a single equation only, were published in [W12], [W13], [PW]; since there we dealt with the initial <u>boundary</u> value problem (with Dirichlet boundary conditions) the exponent q* was smaller than $\frac{n+2(m-1)}{n-2(m+1)}$. The nonlinearities had to fulfill a monotonicity condition (in [W12], [W13]), or some sort or sign condition ([PW]). In [W13] the nonlinearity was also allowed to contain derivatives of u. The exponent $\frac{n+2(m-1)}{n-2(m+1)}-\varepsilon$ with derivatives in the nonlinearity was treated in [W 1]. As already mentioned Goebel [Goe] has treated the initial boundary value problem for a single equation with a monotone nonlinearity f(u) and growth exponent $\frac{n+2(m-1)}{n-2(m+1)}$ and with $n \geq 6m+2$. So far all the papers cited deal with the question of existence of global (in t) classical solutions. The regularity of arbitrary weak solutions was treated in [W16], [W17].

Regularity properties of solutions of single equations and systems $u' + A(t)u + f(u)$ were considered in [W14], [W15]; here f is allowed to grow like $|u|^q$ with $q > 0$ arbitrary, and it may be noted that under monotonicity or sign conditions on f as in [W14], [W15], the existence of a weak solution can be proved. The case $m = 1$ is an exceptional one because of the presence of the weak maximum principle ([LUS, chapter III]); we do not want to go into details here.

An earlier version of theorem V.2.2 was published in [W 9]; the proof was sketched there. The full proof is given in [W 8]. As was already pointed out the application of theorem V.1.1 in the proof of theorem V.2.2 is rather indirect; it's more the method of proof that we need.

VI. References

[ADN I] Agmon, S., Douglis, A. and Nirenberg, L.: Estimates near the boundary for solutions of elliptic partial differential equations satisfying general boundary conditions, I. Comm. Pure Appl. Math. 12, 623-727 (1959).

[ADN II] Agmon, S., Douglis, A. and Nirenberg, L.: Estimates near the boundary for solutions of elliptic partial differential equations satisfying general boundary conditions, II. Comm. Pure Appl. Math. 17, 35-92(1964).

[Ag] Agmon, S.: On the eigenfunctions and on the eigenvalues of general elliptic boundary value problems. Comm. Pure Appl. Math. 15, 119-147(1962).

[B] Bemelmans, J.: Eine Außenraumaufgabe für die instationären Navier-Stokes Gleichungen. Math. Z. 162, 145-173(1978).

[BB] Butzer, P.L., Berens, H.: Semi-Groups of Operators and Approximation. Springer: Berlin, Heidelberg, New York (1967).

[BBW] Butzer, P.L., Berens, H. and Westphal, U.: Representation of Fractional Powers of Infinitesimal Generators of Semigroups. Bull. Am. Math. Soc. 74, 191-196(1978).

[Be] Bergh, J.: A non-linear complex interpolation result. Proceedings of the Conference on Interpolation Spaces and Allied Topics in Analysis, Lund 1983 (M. Cwikel and Y. Peetre ed.). Springer: Lecture Notes in Mathematics 1070, 45-47(1984).

[BL] Bergh, J., Löfström, J.: Interpolation Spaces. Springer: Berlin, Heidelberg, New York (1976).

[Bo] Bogovskij, M.E.: Solution of the first boundary value problem for the equation of continuity of an imcompressible medium. Soviet Math. Dokl. 20, 1094-1098 (1979).

[Br] Browder, F.: On the spectral theory of elliptic differential operators. Math. Ann. 142, 20-130(1961).

[CKN] Caffarelli, L., Kohn, R., and Nirenberg, L.: Partial Regularity of Suitable Weak Solutions of the Navier-Stokes Equations. Comm. Pure Appl. Math. 35, 771-831(1982).

[CW] Chen, C., Wahl, W. von: Das Rand-Anfangswertproblem für quasilineare Wellengleichungen in Sobolevräumen niedriger Ordnung. J. Reine Angew. Math. 337, 77-112(1982).

[Er] Erig, W.: Die Gleichungen von Stokes und die Bogovski-Formel. Diploma-Thesis. Universität Paderborn (1982).

[F] Friedman, A.: Partial Differential Equations. Holt, Rinehart and Winston: New York, Chicago, San Francisco (1969).

[FJR] Fabes, E.B., Jones, B.F., and Riviere, N.M.: The Initial Value Problem for the Navier-Stokes Equations with Data in L^p. Arch. Rat. Mech. Anal. 45, 222-240(1972).

[FK] Fujita, H., Kato, T.: On the Navier-Stokes initial value problem, I. Arch. Rat. Mech. Anal. 16, 269-315(1964).

[Fu] Fujiwara, D.: On the asymptotic behaviour of the Green operators for elliptic boundary problems and the pure imaginary powers of some second order operators. J. Math. Soc. Japan 21, 481-521(1969).

[FuM] Fujiwara, D., Morimoto, H.: An L_r-theorem of the Helmholtz decomposition of vector fields. J. Fac. Sci. Univ. Tokyo, Sect. IA Math. 24, 685-700(1977).

[G] Gerhardt, C.: Stationary Solutions to the Navier-Stokes Equations in Dimension Four. Math. Z. 165, 193-197(1979).

[G1] Giga, Y.: Domains in L_r spaces of fractional powers of the Stokes operator, preprint. To appear in Arch. Rat. Mech. Anal.

[G2] Giga, Y.: Analyticity of the Semigroup Generated by the Stokes Operator in L_r Spaces. Math. Z. 178, 297-329(1981).

[G3] Giga, Y.: Solutions for semilinear parabolic equations in L^p and regularity of weak solutions of the Navier-Stokes system, preprint.

[G4] Giga, Y.: The Stokes operator in L_r spaces. Proc. Japan Acad. 57, 85-89(1981).

[GiM] Giga, Y., Miyakawa, T.: Solutions in L_r to the Navier-Stokes initial value problem, preprint. To appear in Arch. Rat. Mech. Anal.

[Goe] Goebel, R.: Über die Existenz klassischer Lösungen semilinearer parabolischer Differentialgleichungen höherer Ordnung. Math. Z. 184, 511-532(1983).

[GoS] Golovkin, K.K., Solonnikov, V.A.: On the first boundary value problem for nonstationary Navier-Stokes equations. Soviet Math. Dokl. 2, 1188-1191(1981).

[GGZ] Gajewski, H., Gröger, K. und Zacharias, K.: Nichtlineare Operatorgleichungen und Operatordifferentialgleichungen. Akademie Verlag: Berlin (1974).

[H1] Heinz, E.: Beiträge zur Störungstheorie der Spektralzerlegung. Math. Ann. 123, 415-438(1951).

[H2] Heinz, E.: Über die Regularität der Lösungen nichtlinearer Wellengleichungen. Nachr. Ak. d. Wiss. Göttingen, II. Math. Phys. Klasse, 15-26(1975).

[He] Heywood, J.: On Uniqueness Questions in the Theory of Viscous Flows, Acta Math. 136, 60-102(1976).

[HP] Hille, E., Phillips, R.S.: Functional Analysis and Semigroups. Am. Math. Soc.: Providence, R.I., Colloquium Publications 31(1957).

[Hö] Hörmander, L.: Estimates for translation-invariant operators on L_p-spaces. Acta Math. 104, 93-145(1960).

[Ho] Hopf, E.: Über die Anfangswertaufgabe für die hydrodynamischen Grundgleichungen. Math. Nachr. 4, 213-231(1951).

[K1] Kato, T.: Perturbation Theory for Linear Operators. Springer: Berlin, Heidelberg, New York (1966).

[K2] Kato, T.: Nonlinear evolution equations in Banach spaces. Am. Math. Soc.: New York, Proc. Symp. Appl. Math. 17, 50-67(1965).

[K3] Kato, T.: Fractional powers of dissipative operators. J. Math. Soc. Japan 13, 246-274(1961).

[K4] Kato, T.: Fractional powers of dissipative operators, II. J. Math. Soc. Japan 14, 242-248(1962).

[K5] Kato, T.: Strong L^p-Solutions of the Navier-Stokes Equation in $\mathbb{R}^m$, with Applications to Weak Solutions. Math. Z. 187, 471-480(1984).

[Ki1] Kielhöfer, H.: Global Solutions of Semilinear Evolution Equations Satisfying an Energy Inequality. J. Diff. Equations 36, 188-222(1980).

[Ki2] Kielhöfer, H.: Existenz und Regularität von Lösungen semilinearer parabolischer Rand-Anfangswertprobleme. Math. Z. 142, 131-160(1975).

[Kr] Krein, S.G.: Linear Differential Equations in Banach Space. Am. Math. Soc.: Providence, R.I., Translations of Mathematical Monographs 29(1971).

[KS] Kaniel, S., Shinbrot, M.: Smoothness of weak solutions of the Navier-Stokes equations. Arch. Rat. Mech. Anal. 24, 302-324(1967).

[L] Ladyzhenskaja, O.A.: The Mathematical Theory of Viscous Incompressible Flow. Gordon and Breach: New York, London, Paris(1969).

[Ler] Leray, J.: Sur le mouvement d'un liquide visqueux emplissant l'espace. Acta Math. 63, 193-248(1934).

[Li] Lions, J.L.: Quelques Méthodes de Resolution des Problêmes Nonlinéaires. Dunod: Paris(1969).

[LUS] Ladyženskaja, O.A., Ural'ceva, N.N., and Solonnikov, V.A.: Linear and Quasilinear Equations of Parabolic Type. American Math. Soc.: Providence, R.I., Translations of Mathematical Monographs 23(1968).

[M] Masuda, K.: Weak solutions for Navier-Stokes, preprint. To appear in Tohoku Math. J.

[PW] Pecher, H., Wahl, W. von: Klassische Lösungen im Großen semilinearer parabolischer Differentialgleichungen. Math. Z. 145, 255-265(1975).

[R] Rautmann, R.: On optimum regularity of Navier-Stokes solutions at time t = O. Math. Z. 184, 141-150(1983).

[Sb] Sobolevskii, P.E.: Study of the Navier-Stokes Equations by the Methods of the Theory of Parabolic Equations in Banach Spaces. Soviet Math. Dokl. 5, 720-723(1964).

[So1] Sohr, H.: Zur Regularitätstheorie der instationären Gleichungen von Navier-Stokes. Math. Z. 184, 359-376(1983).

[So2] Sohr, H.: Optimale lokale Existenzsätze für die Gleichungen von Navier-Stokes. Math. Ann. 267, 107-123(1984).

[Ser1] Serrin, J.: On the Interior Regularity of Weak Solutions of the Navier-Stokes Equations. Arch. Rational Mech. Anal. 9, 187-195(1962).

[Ser2] Serrin, J.: The initial value problem for the Navier-Stokes equations. University of Wisconsin Press: Madison, Wisconsin, Nonlinear Problems, Proceedings of a Symposium, Madison 1962 (R. Langer ed.), 69-98(1963).

[Sim] Simader, C.G.: Remarks on the truncation method for functions in Sobolev spaces. Preprint.

[Sol1] Solonnikov, V.A.: Estimates for solutions of non-stationary Navier-Stokes equations. J. Soviet Math. 8, 467-529(1977).

[Sol2] Solonnikov, V.A.: Estimates of the solutions of a nonstationary linearized system of Navier Stokes equations. American Math. Soc. Transl. 75, 1-116(1968).

[SOW1] Sohr, H., Wahl, W. von: On the Singular Set and the Uniqueness of Weak Solutions of the Navier-Stokes Equations. Manuscripta math. 49, 27-59(1984).

[SOW2] Sohr, H., Wahl, W. von: On the Regularity of Weak Solutions of the Equations of Navier-Stokes. Preprint.

[Ta] Tanabe, H.: Equations of Evolution. Pitman: London, San Francisco, Melbourne (1979).

[Tem1] Témam, R.: Navier-Stokes Equations. North Holland: Amsterdam, New York, Oxford (1977).

[Tem2] Témam, R.: Behaviour at time t = 0 of the solutions of semilinear evolution equations. J. Diff. Equations 43, 73-92(1982).

[Tr1] Triebel, H.: Interpolation Theory, Function Spaces, Differential Operators. North Holland: Amsterdam, New York, Oxford (1978).

[Tr2] Triebel, H.: Spaces of Besov-Hardy-Sobolev Type. Teubner: Leipzig. Teubner Texte zur Mathematik 15 (1978).

[W1] Wahl, W. von: Analytische Abbildungen und semilineare Differentialgleichungen in Banachräumen. Nachr. Ak. d. Wiss. Göttingen, II. Mathem. Phys. Klasse, 153-200(1980).

[W2] Wahl, W. von: Über das Verhalten für $t \to 0$ der Lösungen nichtlinearer parabolischer Gleichungen, insbesondere der Gleichungen von Navier-Stokes. Bayreuther Math. Schr. 16, 151-277(1984).

[W3] Wahl, W. von: On Nonlinear Evolution Equations in a Banach Space and on Nonlinear Vibrations of the Clamped Plate. Bayreuther Math. Schriften 7, 1-93 (1981).

[W4] Wahl, W. von: Regularitätsfragen für die instationären Navier-Stokesschen Gleichungen in höheren Dimensionen. J. Math. Soc. Japan 32, 263-283(1980).

[W5] Wahl, W. von: Klassische Lösbarkeit im Großen für nichtlineare parabolische Systeme und das Verhalten der Lösungen für $t \to \infty$. Nachr. Ak. d. Wiss. Göttingen, II. Mathem.-Phys. Klasse, 131-177(1981).

[W6] Wahl, W. von: Gebrochene Potenzen eines elliptischen Operators und parabolische Differentialgleichungen in Räumen hölderstetiger Funktionen. Nachr. Ak. d. Wiss. Göttingen, II. Math.-Phys. Klasse, 231-258(1972).

[W7] Wahl, W. von: Regularity Questions for the Navier-Stokes equations. Proceedings IUTAM Symp. Paderborn 1979 (R. Rautmann ed.). Springer Lecture Notes 771, 538-542(1980).

[W8] Wahl, W. von: Regularity of Weak Solutions of the Navier-Stokes Equations. Proc. Symp. Pure Math., AMS Summer Institute on Nonlinear Functional Analysis and Applications, Berkeley 1983 (F.E. Browder ed.). Am. Math. Soc.: Providence, R.I. To appear.

[W9] Wahl, W. von: Nichtlineare Evolutionsgleichungen. Teubner: Leipzig. Teubner Texte zur Mathematik 50, 294-302(1983).

[W10] Wahl, W. von: Instationary Navier-Stokes Equations and Parabolic Systems. Pac. J. Math. 72, 557-569(1977).

[W11] Wahl, W. von: The equation $u' + A(t)u = f$ in a Hilbert space and L^p-estimates for parabolic equations. J. London Math. Soc. (2)25, 483-497(1982).

[W12] Wahl, W. von: Lineare und semilineare parabolische Differentialgleichungen in Räumen hölderstetiger Funktionen. Abh. Math. Sem. Hamburg 43, 234-262(1975).

[W13] Wahl, W. von: Semilinear elliptic and parabolic equations of arbitrary order. Proc. Royal Soc. Edinburgh 78A, 193-207(1978).

[W14] Wahl, W. von: Semilineare parabolische Differentialgleichungen mit starker Nichtlinearität. Manuscripta math. 16, 395-406(1975).

[W15] Wahl, W. von: Regularitätssätze für semilineare parabolische Differentialgleichungen mit starker Nichtlinearität. Nachr. Ak. d. Wiss. Göttingen, II. Math. Phys. Klasse, 73-82(1976).

[W16] Wahl, W. von: Regularity of Weak Solutions to Parabolic Equations of Arbitrary Order. J. London Math. Soc. (2)15, 297-304(1977).

[W17] Wahl, W. von: On the extension of a result of Ladyženskaja and Ural'tseva for second-order parabolic equations to equations of arbitrary order. Ann. Pol. Math. XLI, 63-72(1983).

[We] Weissler, F.B.: The Navier-Stokes Initial Value Problem in L^p. Arch. Rat. Mech. Anal. 74, 219-230(1981).

[Y1] Yosida, K.: On the differentiability of semigroups of linear operators. Proc. Japan Acad. 34, 337-340 (1958).

[Y2] Yosida, K.: Functional Analysis. Springer: Berlin, Heidelberg, New York (Fifth edition, 1978).